Practical NMR Spectroscopy Laboratory Guide: Using Bruker Spectrometers

Practical NMR Spectroscopy Laboratory Guide: Using Bruker Spectrometers

John S. Harwood
Department of Chemistry, and Purdue Interdepartmental NMR Facility
Purdue University West Lafayette, IN, USA

Huaping Mo
Department of Medicinal Chemistry and Molecular Pharmacology, and
Purdue Interdepartmental NMR Facility Purdue University
West Lafayette, IN, USA

AMSTERDAM • BOSTON • HEIDELBERG • LONDON
NEW YORK • OXFORD • PARIS • SAN DIEGO
SAN FRANCISCO • SINGAPORE • SYDNEY • TOKYO

ELSEVIER   Academic Press is an imprint of Elsevier

Academic Press is an imprint of Elsevier
125, London Wall, EC2Y 5AS.
525 B Street, Suite 1800, San Diego, CA 92101-4495, USA
225 Wyman Street, Waltham, MA 02451, USA
The Boulevard, Langford Lane, Kidlington, Oxford OX5 1GB, UK

Notices
Knowledge and best practice in this field are constantly changing. As new research and
experience broaden our understanding, changes in research methods or professional practices,
may become necessary.

Practitioners and researchers must always rely on their own experience and knowledge in
evaluating and using any information or methods described herein. In using such information or
methods they should be mindful of their own safety and the safety of others, including parties for
whom they have a professional responsibility.

To the fullest extent of the law, neither the Publisher nor the authors, contributors, or editors,
assume any liability for any injury and/or damage to persons or property as a matter of products
liability, negligence or otherwise, or from any use or operation of any methods, products,
instructions, or ideas contained in the material herein.

ISBN: 978-0-12-800689-4

British Library Cataloguing-in-Publication Data
A catalogue record for this book is available from the British Library

Library of Congress Cataloging-in-Publication Data
A catalog record for this book is available from the Library of Congress

For Information on all Academic Press publications
visit our website at http://store.elsevier.com/

This book has been manufactured using Print On Demand technology.

Working together
to grow libraries in
developing countries

www.elsevier.com • www.bookaid.org

CONTENTS

ACKNOWLEDGMENTS

The authors would like to thank their families and co-workers for their patience and support during the time this book was being worked on.

The authors would also like to thank the following Purdue students for their assistance in the development of the book's material:

Jinshan Gao, Lingyan Liu, Oscar Morales, Panuwat Padungros, Tairan Yuwen, Anura Indulkar, Chad Keyes, Mike Mazzotta, and Sarah St. John.

Finally, we would like to acknowledge several helpful conversations with Dr Josh Kurutz during the early stages of this project.

OVERVIEW

This book is an adaptation of a manual written for a laboratory course in practical solution-state NMR spectroscopy given at Purdue University by the authors in the summer of 2010. It is designed to help provide nonexpert NMR users, typically graduate students in chemistry and related disciplines, an introduction to various facets of NMR spectroscopy which we expect will be useful to them during their time in graduate school and beyond. This book will lead the reader through a series of hands-on exercises using an NMR spectrometer which are designed to introduce various NMR concepts and experiments and give the reader experience in running these experiments. It is hoped that this work will be useful both as a text for instructors of a practical NMR course and also as a reference for spectrometer administrators or NMR facility directors when doing user training. It is also hoped that this work will be of use to graduate students working on their own (with assistance available) to expand their experience with the NMR spectrometer.

Therefore, this book assumes that readers are capable of operating the NMR spectrometer, and that they are able to obtain at least standard ^1H and ^{13}C NMR spectra of a typical organic sample without external assistance or the use of automation. The nominal skills needed to do this would include creating and moving between datasets, changing, locking, and shimming the sample, loading parameters, setting up the experiment and acquiring the data, and processing the data and plotting the resulting spectra. This book assumes further that the reader has had some experience with the interpretation of ^{1}H and ^{13}C NMR spectra of organic samples and at least an introduction to the basic theories of NMR spectroscopy.

Since the authors' course was originally designed to fit into an 8-week summer session, there were 7 laboratory sessions written, with the expectation that the last week would be used for self-directed study in the form of a structure determination project. This structure determination project was used as part of the final examination for the class. For this book we have added an additional laboratory session. Instructors wishing to fit this material into an 8- or 16-week semester may do so as they see fit. The laboratory sessions are set up so that each one should be able to be completed (with most spectrometers) within two laboratory meetings of 3 hours each.

For the course textbook we used *High-Resolution NMR Techniques in Organic Chemistry*, by Timothy D.W. Claridge (2nd edition, Elsevier, Oxford, 2009). We will note the relevant chapters in this edition of Claridge at the beginning of each chapter of the current work.

In order to limit the scope of the book we will cover only solution-state NMR spectroscopy and limit ourselves to ^{1}H- and ^{13}C-observe NMR.

NMR SAMPLES REFERENCED IN THE TEXT

Throughout the course of this book we will describe experiments using in total 13 different samples. We will list them here and abbreviate them as samples A−M. At the beginning of each lab we will list the suggested samples using these abbreviations.

Sample	Description
A	Quinine HCl in CDCl$_3$, ca. 40 mg/mL (1)
B*	CHCl$_3$ in acetone-d6 (lineshape standard sample), concentration 0.3−5% v/v (2)
C*	Ethylbenzene in CDCl$_3$, 0.1% v/v (^{1}H sensitivity standard sample)
D*	Ethylbenzene in CDCl$_3$, 10% v/v (^{13}C sensitivity standard sample)
E	CHCl$_3$ in CDCl$_3$, ca. 20% v/v (1)
F	H$_2$O in D$_2$O, ca. 20% v/v, between two teflon vortex plugs ca. 1 cm apart (3)
G*	5 mM lysozyme in 90% H$_2$O/10% D$_2$O (4)
H*	1 mM ^{13}C, ^{15}N-labeled ubiquitin in 90% H$_2$O/10% D$_2$O (4)
I	Water in octanol (5)
J	Octanol in water (5)
K*	100% methanol (temperature calibration standard sample)
L*	0.1 mg/mL GdCl$_3$ in D$_2$O (doped D$_2$O) (6)
M	Diffusion (DOSY) sample (1, 7)

Notes:
*Sample is commercially available.
(1) We recommend that these samples be at least 700 μL in volume and that the tubes be sealed.
(2) The correct concentration will depend on the ^{1}H sensitivity of the spectrometer/probe combination you are using.
(3) A more detailed description of this sample is given in Chapter 5. If available a Shigemi™ tube may be used instead of a conventional tube with vortex plugs and will give a better result.
(4) Different, locally-available, samples may be used instead of the commercial samples. If you do not do the biomolecular-NMR experiments in Chapter 6 these samples will not be needed.
(5) These samples will be discussed in Chapter 7 in the section on quantitation by NMR.
(6) A sample of plain D$_2$O, with or without a small amount of H$_2$O, will work.
(7) The sample we employed for DOSY experiments is discussed in Chapter 8.

Many of these samples are standard samples used for NMR spectrometer performance testing, and as such will typically be available as part of any NMR facility's testing equipment. Other samples will need to be made up specifically for these exercises. Regarding sample A, the combination of the quinine HCl sample concentration and the spectrometer's ^{13}C sensitivity should allow the acquisition of the quinine ^{13}C spectrum in a reasonable amount of time for your own laboratory arrangements. On a spectrometer with a ^{13}C sensitivity specification of ca. 200:1, a 40 mg/mL sample will provide a good-quality ^{13}C spectrum in 10−15 min. Higher concentrations may be used if necessary since quinine HCl is freely soluble in chloroform.

Throughout the book we expect the student to use many of the spectra run in the labs to ultimately obtain the full ^1H and ^{13}C chemical shift assignments for quinine. In this fashion we hope to give the student useful experience in the assignment process and, in the choosing and use of various NMR experiments, to facilitate that process. We picked quinine because it is readily available at low cost and produces, at least in our opinion, interesting and pedagogically useful spectra. Other samples were employed for certain experiments because we felt that they provided useful example spectra for the experiments being presented. Instructors are of course free to use any sample(s) of their own choosing.

SPECTROMETER REQUIREMENTS

When the material for this book was first being generated, the laboratory sessions were worked out using a Bruker Avance DRX500 NMR spectrometer equipped with a Linux computer running TopSpin 1.3 pl8. This spectrometer was equipped with either a 5-mm TXI Z-gradient cryoprobe or a 5-mm TBI Z-gradient conventional probe. Further checking of the material was done using both a second Avance DRX500 spectrometer running Linux TopSpin 1.3 pl8 and equipped with a 5-mm BBFO Z-gradient probe, as well as an AV-III-800 NMR spectrometer running Linux TopSpin 2.1 pl6 and equipped with a 5-mm Z-gradient TXI probe. Given the variability of the NMR hardware and software versions extant, combined with the possible variations of host computer operating systems, we would recommend that any instructor or spectrometer/facility administrator check through these experiments in detail prior to using them in a class, in case there are any incompatibilities between what we have written and a specific hardware/probe/software/operating system combination. For example, we have not used spectrometers with Windows-based host computers in the preparation of this work. Furthermore, the possibility of an error or omission on the part of the authors cannot be completely discounted!

The procedures in this book were written based on the assumption that samples in conventional 5 mm NMR tubes will be used. The only specific spectrometer configuration requirements are that Z-axis field gradient hardware will be needed for the exercises in Chapters 5–8, and an "inverse" geometry probe (e.g., QXI, TXI, TBI, BBI or a SmartProbe™) is recommended to get the best results for the experiments in Chapter 6.

ACTIVITIES

The activities in each chapter will be described using step-by-step instructions to guide the user through all of the procedures. As the book progresses we will reduce the level of detail to correspond with the reader's increasing experience. When digressions from the main flow of the activities are necessary, for example, to explain or provide background information for a specific procedure, these will be described in an appendix for that chapter so as not to interrupt the progression of the laboratory. At the end of each chapter we will include a few questions which we have asked the students to complete as part of their lab write-ups when the authors gave the course. Literature references will also be included when relevant.

We will use some typeface changes to denote the following:

1. The geneva font with extra spacing will be used for typed commands, e.g., type zg to acquire data.
2. ALL CAPS will be used to denote buttons or inputs to be clicked, e.g., click WOBB-SW to change the wobb sweep width.
3. *ALL CAPS ITALICS* will be used for instrument parameters as entered in menus, e.g., *PL1* = 0 dB. Note that the parameter can be queried or changed through the command line by using lower case, e.g., pl1 (return).
4. Menu titles and selections therein will be in quotation marks, with sequential choices denoted with a right-pointing arrow, e.g., "Spectrometer" → "Shim control" menu.

For this edition of the guide we will use Bruker-specific nomenclature when applicable. For ease of description, we use typed commands to drive most operations, though the operations can usually be carried out also through the use of menu pulldowns, icon clicks, or other methods.

It should be noted that the setting up of experiments and the parameter inputs to do so are written with the order of the experiments in mind. That means that if a certain parameter has been set during an earlier experiment it may not be listed again for input for the current experiment unless its value needs to be changed. Therefore, if instructors wish to carry out experiments in an order different than what is shown here, they should confirm the input of all the relevant parameters needed for a given experiment.

REFERENCE

Claridge, T.D.W. (Ed.)., 2009. High-Resolution NMR Techniques in Organic Chemistry, second edition. Tetrahedron Organic Chemistry Series, Vol 27. Elsevier, Oxford.

CHAPTER *1*

Basics and Spectrometer Performance Checks

OVERVIEW

This chapter's exercises are designed to introduce some of the fundamentals of NMR data acquisition and processing. We will look into some of the details of shimming, pulse calibration, sensitivity checking, and data acquisition and processing. The material covered in this chapter is in Chapters 2 and 3 of the Claridge book.

Note: For each laboratory, we recommend that you make a new experiment name such as "chapter-1" and use a new experiment number (expno) for each experiment you run. Don't lose data by accidentally overwriting it! If in doubt always run an experiment in a new expno.

SAMPLE AND SPECTROMETER REQUIREMENTS

This chapter's exercises will use samples A, B, C, and D. Since the NMR experiments in this chapter are simple 1D acquisitions, there are no special requirements for the spectrometer hardware.

ACTIVITIES

Quinine

Part 1—Familiarization

1. Set up (using different experiment numbers) ^{1}H and ^{13}C datasets using your spectrometer's standard parameter sets and procedures (discussion and examples of the standard parameter sets used at in the authors' facility are presented in Appendix 1.1).

Practical NMR Spectroscopy Laboratory Guide: Using Bruker Spectrometers.
DOI: http://dx.doi.org/10.1016/B978-0-12-800689-4.00001-X

2. Load the quinine/$CDCl_3$ sample into the magnet and complete the locking and shimming process. Tune the probe for both ^1H and ^{13}C (command wobb or cwobb or atmm/atma—for a discussion of probe tuning, see Appendix 1.2). If you are not experienced with probe tuning please obtain assistance from the spectrometer administrator for this step. Note that we will expect that probe tuning be done whenever a sample is changed or a new nucleus observed, even if it is not explicitly mentioned in the text.

3. Acquire the data and process the spectra using manual phasing, and check the chemical shift referencing. Edit the title text (Title tab or setti) to reflect the experiment and sample used.

4. Plot both spectra, including peak-picking and integrations for the ^1H. Check that the integration values look reasonable.

5. Note the effect of exponential multiplication by processing your data using ef versus ft. Try different values of lb to assess the impact of differing exponential multiplication functions upon the spectrum.

6. Using Figure 1.1 as a guide, try processing the ^1H data using a couple of different window functions, such as Gaussian and sine-bell. To select one of these use the wm command. In the popup window use the pulldown to select the desired window function, then enter the necessary controlling parameters in the appropriate fields: use gb and lb for Gaussian and ssb for sine-bell. Starting values for these parameters would be as follows:

gb	0.3 (range is 0–1)
lb	−0.5 (approximate value is ca. − (1/AQ))
ssb	1 or 2 (1 for sine-bell, 2 for cosine)

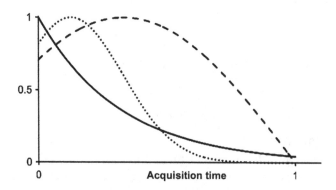

Figure 1.1 Three frequently used data processing window functions plotted with the horizontal axis representing the unit acquisition time: exponential (solid line); Gaussian-exponential (dotted line); and a 45° shifted sine-bell (dashed line).

To apply the window function to the raw NMR data (the Free Induction Decay or FID), click the OK button in the popup window. Then to generate the new spectrum use either the ft or fp command (fp applies the previously-determined phase correction, using processing parameters *PHC0* and *PHC1*, to the spectrum, whereas ft generates a spectrum without phase correction). Feel free to experiment with other window functions—the original FID remains unchanged, only the spectrum changes.

7. Print out at least three spectra obtained with different processing functions. Show the areas of the spectra where the processing function has the most impact.

8. Examples of the ^{1}H and ^{13}C spectra of quinine at 500/125 MHz are shown in Appendix 1.3.

Part 2—Shimming and Lineshape

1. Replace the quinine sample with the $CHCl_3$/acetone-d6 sample. Before locking, note the signal in the lock display: why is there no lock signal wiggle apparent?

2. Create a new dataset (edc or new command). Lock on acetone-d6. Make sure the sample spinning is on.

3. Shim Z1, Z2, and Z3 by hand (using either the BSMS physical keypad or the command bsmsdisp to display the software shimming interface) to get the best lock level. Keep the lock power low to avoid saturating the lock signal (a saturated lock signal may drift up and down even when shims are not being changed, and may be less responsive than normal to shim changes).

4. Run a ^{1}H spectrum in the new dataset using standard parameters. Contrary to normal practice, for this spectrum set the chemical shift of the $CHCl_3$ peak to 0 and change the displayed axis units from ppm to Hz (h/p icon in the upper icon bar).

5. Copy the data to a new expno, using either wrpa and re commands (e.g., wrpa 3 followed by re 3 to duplicate the current dataset to experiment number 3 and then to load it) or the new command. Optimize the *O1*, *SW*, and *TD* parameters to give a data (FID) acquisition time (*AQ*) of ca. 10 s—this ensures adequate digital resolution in the resulting spectrum. Set, or check that, the *SI* parameter equal to *TD/2*. Rerun the spectrum and take a close look at the $CHCl_3$ peak. Expand the spectrum ca. ± 30 Hz around the $CHCl_3$ peak and define your expanded region as the plot region. Using the cursor left-click and drag mode, measure the linewidth

and lineshape by hand, at ½-height (for linewidth) and at 0.55% and 0.11% of the total height (for lineshape, or "hump") of the line. Make a note of your results. Then type humpcal to use the au program to make the same measurements using a peak-fitting routine. Does it meet specification? Note that the lineshape specifications are specific to the instrument and probe you are using, so you will need to check the instrument documentation or with the spectrometer administrator to find this information. If neither is readily available, reasonable values for a conventional probe would be a ½-height linewidth of less than 0.8 Hz, and a lineshape of less than 7.0 and 14.0 Hz for the 0.55% and 0.11% measurements, respectively. For making the linewidth measurements by hand it can be advantageous to set the cy equal to 100 or 1000, whereby, for example, the linewidth at 0.55% height would be measured when the cursor height reading is 0.55 or 5.5 cm, respectively. Be aware that if the zoom region being displayed is too narrow, the cursor may not display the correct height value for its location. Usually this can be remedied by broadening the displayed zoom region (i.e., alter the zoom display to include more Hz). An example of the lineshape measurement is shown in Figure 1.2.

6. Work on the shimming to improve the lineshape. If the lineshape is asymmetric, Z2 and Z4 need work. Check also Z3 and Z5. If Z4 is altered re-shim Z1, 2, and 3—often a good strategy is to alter Z4 by 500 units, re-shim, and then observe the spectrum to see if the lineshape has improved. If it has, repeat the process by changing Z4 another 500 units in the same direction until the optimum lineshape is obtained. If the lineshape is worse, change Z4 back 500 units in the other direction from the first change, re-shim and reacquire. If Z5 is altered re-shim Z1 and Z3, then Z2. Some of the Z-axis (spinning) shim functions' effects upon the lineshape are shown in Figure 1.3.

7. Reacquire spectra as needed.

8. Turn off sample spinning and optimize the low-order nonspinning shims as follows:
   a. Z1, X, XZ, XZ2
   b. Z1, Y, YZ, YZ2
   c. XY, X2 − Y2, XZY
   d. X, Y.
   Note that on some spectrometers (equipped with "matrix" shims) the *physical* keypad operation for nonspinning shims requires selection of

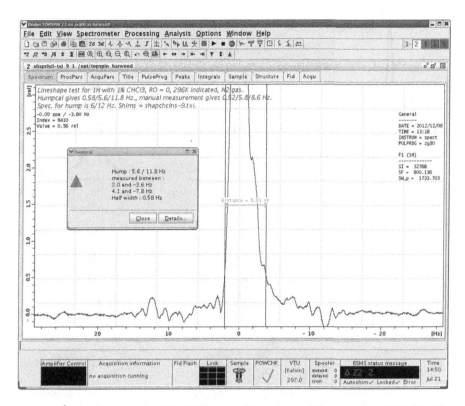

Figure 1.2 ^1H lineshape test of a 0.1% CHCl$_3$/acetone-d6 sample recorded on a Bruker AV-III 800 MHz spectrometer equipped with TXI probe at 298 K. The nonspinning linewidth at ½-height, 0.55% and 0.11% height is measured using humpcal as 0.58, 5.6, and 11.8 Hz. The manual measurement of the linewidth and ca. 0.55% height is shown as well. Note the "Value = 0.56 rel" text displayed in the cursor information area on the left-side of the spectrum window. This "Value" is the height of the datapoint at the cursor's location relative to the peak height (cy) which was set to 100 cm. When measuring lineshape by hand you may not be able to get a height of exactly 0.55%; just use a very close height or interpolate between adjacent data points. Low-frequency (~10 Hz) noise from floor vibration is seen at the base of the peak.

the X, Y, XY, or X2 − Y2 button followed by the required Z function. For example, to adjust the X shim press the X button followed by the Z0 button; to adjust the YZ2 shim press the Y button followed by the Z2 button, etc. Some of the effects of the nonspinning shims upon the lineshape are shown in Appendix 1.4.

9. Restart spinning and touch up Z1, Z2, and Z3. Reacquire. Decide if your result is good enough, if it is not then repeat the shimming process again.

10. Acquire your last, best, spectrum with spinning on. Process and plot. Include your lineshape results and spinning status in the title of the printout. Did you achieve a lineshape which met specifications?

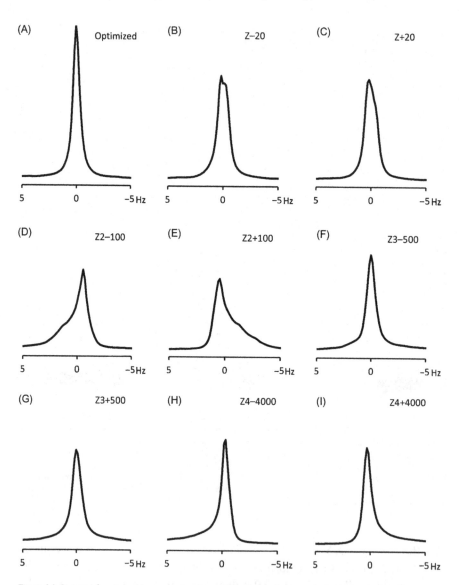

Figure 1.3 Impact of various shims upon the observed lineshape of the residual solvent signal from a CDCl₃ sample obtained on a Bruker Avance DRX 500 MHz spectrometer equipped with a BBO probe and no sample spinning. The various shim conditions are: optimized (A); Z shim is reduced (B) or increased (C) by 20; Z2 shim is reduced (D) or increased (E) by 100; Z3 shim is reduced (F) or increased (G) by 500; Z4 shim is reduced (H) or increased (I) by 4000.

11. Turn spinning off and reacquire the spectrum in a new expno. Print this with the lineshape information and spinning status again in the title.
12. Save your shim settings using the wsh command (e.g., wsh chapter1-acet.shm).

Part 3—Pulse Calibration and Sensitivity for ^{1}H

1. In order for the spectrometer to obtain its highest sensitivity several things have to be optimized. Put the 0.1% ethylbenzene sample in the magnet, spin, lock on $CDCl_3$, and shim. Use a new expno. Tune the probe for ^{1}H. Run the ^{1}H spectrum using standard parameters.

2. Now set the pulse program (*PULPROG*) to zg, $NS = 1$, $DS = 0$, and $P1 = 3\,\mu s$. Check that the transmitter power level (parameter *PL1*) is the correct value for your spectrometer. If this information is not readily available a power level of 0 dB will usually be acceptable. Check the receiver gain using rga. Rerun and phase.
   Note: In the Bruker convention, a more positive power (dB) level signifies less RF power, e.g., 0 dB is a typical moderately-high power while 120 dB is very low (close to zero) power. Thus the *PL* value is referring to an attenuation (power-reduction) level from a reference power level.

3. Copy data and use a new expno. We will now determine the 90-degree pulse (p90) length using the 360-degree method. A 360° pulse will give almost zero signal, with less than 360° giving a weak negative peak and more than 360° giving a weak positive one. Enter a value for *P1* that is about four times the specified p90 value for the probe you are using (a typical ^{1}H high power pulse length is expected to be in the order of 10 μs). Rerun and apply the existing phase correction (efp). Adjust the *P1* value and reacquire the data until you get a spectrum with almost zero signal intensity. The p90 value will be equal to this *P1* divided by 4. There is also an automated routine available for carrying out this process called paropt, the use of this is presented in Appendix 1.5 (Figure 1.4).

4. Copy and use a new expno. For *P1* enter the value you determined above for the p90, and run one scan after checking the receiver gain setting with rga. Carefully phase the resulting spectrum.

5. New expno. Adjust the acquisition and processing parameters as follows: $D1 = 60$ s, $O1P = 4$ ppm, $SW = 10$ ppm, $TD = 32K$ (i.e. 32768 points), $LB = 1.0$, and $SI = 16K$. Be sure to have the calibrated parameters for p90 (*P1*) entered. Rerun one scan, process using ef, and determine the signal-to-noise ratio. This is most easily done through the use of the sinocal program. Type sinocal to start the routine, which will then ask for several inputs. The signal region is chosen to be the methylene quartet, defined between 2.8 and 2.5 ppm; the noise region is defined between 7.1 and 2.8 ppm; the noise width is entered as the ppm equivalent of 200 Hz (i.e., 0.4 ppm at 500 MHz). Print this spectrum with the s/n value included in the title text (Figure 1.5).

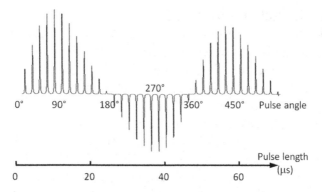

Figure 1.4 A 360° pulse calibration that is based on observation of the signal after variable length on-resonance pulse excitations. The 360° pulse is the second null point: half of that leads to first null while ¼ or ¾ of 360° leads to the maximum positive or negative signal.

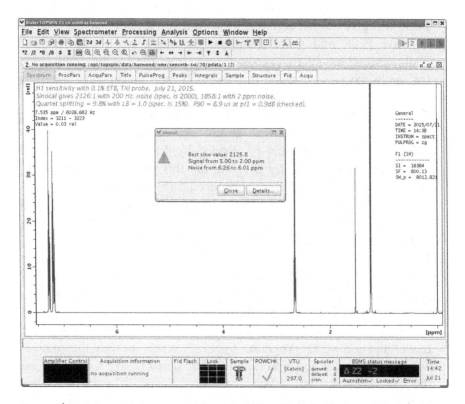

Figure 1.5 ^1H signal-to-noise ratio measured as 2150:1 for a 0.1% ETB sample (20 Hz spinning) using a Bruker AV-III 800 MHz spectrometer equipped with a TXI probe at 297 K. This ratio is calculated as the methylene peak height compared to the twice the RMS noise level of a 200-Hz-wide band of random noise in the spectrum.

6. OPTIONAL. Obtain your spectrometer administrator's permission before carrying out steps 7–9.
7. Create a new dataset in a new expno. Using the probe tuning routine, detune the probe so that the tuning dip is moved away from the correct ^1H frequency by ca. 1 MHz.
8. Repeat the p90 calibration procedure.
9. New expno. Repeat the sensitivity-test scan using the new p90 calibration value, run `sinocal` on the resulting spectrum and print the spectrum out. Is this result different from that obtained in part 5? Why or why not?

Part 4—Pulse Calibration and Sensitivity for ^{13}C

1. Put in the 10% ETB sample, spin, lock on $CDCl_3$, and shim. Set up a new expno.
2. Run the ^1H spectrum to check that the shimming is acceptable.
3. Set up a ^{13}C acquisition using the standard parameters. We assume here that the *PL1* value in the standard parameters is correct.
4. Check the probe tuning for both ^1H and ^{13}C.
5. Acquire a ^{13}C spectrum with 8 scans. Check that it looks OK.
6. Copy the data to a new expno and set $DS = 0$, $NS = 1$, $PULPROG$ = zgpg and $P1 = 5\,\mu s$. Run this and phase the spectrum carefully.
7. Determine the ^{13}C p90 in the same fashion as in Part 3. Use a long-enough D1 delay (30–60 s) between acquisitions so that the signal doesn't get saturated.
8. Copy data to a new expno and set up the parameters for the ^{13}C sensitivity check with ^1H decoupling as follows:

PULPROG	zgdc
RG	maximum value (or close to it) for your spectrometer
D1	300 s
O1P	80 ppm
SW	200 ppm
P1	Correct p90 value
TD	256K
O2P	ca. 5 ppm (i.e., the approximate center of the ETB ^1H spectrum)
CPDPRG2	waltz16
PCPD2	Use the value from your spectrometer's standard ^{13}C parameters
PL12	Use the value from your spectrometer's standard ^{13}C parameters
LB	0.3
SI	128K

Note: The *PCPD2/PL12* parameter combination is the calibration for a ^{1}H p90 of ca. 80 µs (500 MHz spectrometer) to 60 µs (800 MHz). This pulse is used as part of the waltz16 composite-pulse-decoupling sequence (*CPDPRG2*) to give uniform ^{1}H decoupling across the ^{1}H spectrum width.

Rerun one scan, process, and determine the signal-to-noise ratio. Use the `sinocal` routine with the following inputs: signal region between 130 and 125 ppm; noise region defined between 125 and 29 ppm; noise width of 40 ppm. Print this spectrum with the s/n value included in the title text.

Part 5—Viewing and Editing Text Files in TopSpin

1. Macros, au programs, and pulse programs are examples of types of text files that you will work with as a normal part of spectrometer operation. The `sinocal` program we used for the signal-to-noise ratio calculation is an example of an au program. Often it is useful to be able to view or even edit your own copies of these types of files. In the following steps we will take a look at each one of these types of files.

2. Macros are files which contain a series of TopSpin commands to be executed sequentially. Type `edmac` to open a menu of macros which are included with the software. We can edit our own macro as an example, in this case to acquire data and process it automatically.

   Type `edmac zgefp.xyz` (where xyz = your initials) followed by the return key to enter the editing mode for the new macro titled "zgefp.xyz." In the editing window enter the following as shown:

   rga

   zg

   ef

   apk

   Terminating the editing process is slightly different in different versions of TopSpin. For TopSpin 1.3, click the OK button at the bottom left of the edmac window (or type alt-o) to save your macro and exit the editor. For TopSpin versions 2, go to the File pulldown and select Close, then confirm the save request in the popup window to exit the editor. Whenever you want to execute this macro simply type `zgefp.xyz` and the series of commands entered in the macro will be carried out.

3. Au programs are programs that can operate on the spectrometer or on data. Here we will create a version of an au program that we

use at our facility called "standard." This is used to set the spectrometer up to be left with the standard CDCl₃ sample in the magnet after someone has finished using the spectrometer. It assumes that a ¹H dataset is loaded as the current dataset. Type edau standard.xyz (again, xyz = your initials) to open the au program editor with the new file named standard.xyz. In the editor window enter the following text:

```
/*file location: /opt/topspin/exp/stan/nmr/au/src/standard.xyz */
/*this program will read shim file shims.probe and lock on CDC13
    solvent */
/*this program should be executed from a 1H dataset which DOES NOT
    contain critical data*/
```

```
char solvent[20];                        /*initialize the character variable named 'solvent' */
GETCURDATA                               /*load the current dataset */
FETCHPAR("SOLVENT", solvent)             /*save the current solvent value into 'solvent' */
STOREPAR("SOLVENT", "CDCl3")             /*set the solvent variable to CDC13 */
RSH("shims.probe")                       /*load the shims from file 'shims.probe' */
II                                       /*initialize interfaces */
LOPO                                     /*set up lock for new solvent */
LOCK                                     /*lock on CDC13 */
STOREPAR("SOLVENT", solvent)             /*set solvent parameter back to original value */
QUIT                                     /*finish the au program */
```

Note that the text between the /* and */ symbols is included as commentary to help explain the operation of this au program, so it does not have to be included when editing your au program. If you do include it in your program it will be treated as commentary and will not be compiled as part of the program. If you do include comments be sure to include the designating symbols (/* */) correctly. Terminating the editing process is slightly different in different versions of TopSpin. For TopSpin 1.3, after entering all the lines, click the SAVE, EXIT, and COMPILE button at the bottom of the editor window (or type alt-x) to save and compile the program and exit the editor window. For TopSpin versions 2, first click the COMPILE button to compile the program, then go to the File pulldown and select Close, then confirm the save request in the popup window to exit the editor. Assuming the compilation completes successfully, to execute this program simply type standard.xyz.

4. Type edpul to open a menu of all the pulse programs on the spectrometer. Choose, for example, "zg" click the EDIT button at the bottom of the window to open it. You will now see the syntax of a simple pulse program. In addition, the PulseProg tab in the spectrum window will show you the pulse program in the

current experiment. We will edit a pulse program as part of Chapter 3 activities.

SUMMARY

The lab report should include the following:

1. Quinine: Three ^{1}H spectra, each processed with a different window function, and the ^{13}C spectrum.
2. 0.3% CHCl$_3$ in acetone-d6: Lineshape spectra with and without sample spinning, with results in the title text.
3. 0.1% ETB sample: ^{1}H spectrum, spectrum showing ca. 360° pulse, sensitivity spectrum with results in the title text.
4. 10% ETB sample: Standard ^{13}C spectrum, sensitivity spectrum with results in the title text.
5. Suggested questions for consideration:
   a. What is the significance of the 90° pulse?
   b. When you put the lineshape sample into the magnet, why was there not a lock wiggle signal in the lock window as there was when changing to another CDCl$_3$ sample?
   c. Why does the residual solvent peak in your first lineshape-sample spectrum look like it does? Why does it show its specific multiline pattern?
   d. Can you make any preliminary assignments for the ^{1}H and ^{13}C spectra of quinine?
   e. If you measured the ^{1}H sensitivity after detuning the probe (OPTIONAL), what accounts for the difference between the two results, and why? Do you notice any lineshape change correlating with a sensitivity change?

APPENDIX 1.1 STANDARD PARAMETER SETS

In the context of this work, when we use the term "standard parameters" we are referring to the parameters used to obtain a routine 1D ^{1}H or ^{13}C spectrum on a typical organic sample. Thus, this is referring to the most basic type of NMR experiment setup. The parameters in these experiments relevant to the NMR data acquisition are the RF pulse width and pulse power, the precise spectrometer frequency, or transmitter offset, the sweep width, the number of data points acquired, and the interpulse (or relaxation) delay. In the case of ^{13}C acquisition, there are parameters for ^{1}H decoupling including the

RF power, the decoupling pulse width, the decoupling pulse program, and the transmitter offset for the decoupling channel. The parameter sets also include the required data processing parameters, such as the type of window function applied to the FID and the settings for that function, the number of real data points in the final spectrum, and the chemical shift referencing information.

The Bruker TopSpin software includes a large number of parameter sets that usually will be installed as part of the software installation when the spectrometer is first set up. The parameter sets for routine 1D ^{1}H or ^{13}C observation are named PROTON and C13CPD, respectively. However, these parameter sets will not work as is because by default they do not include the relevant pulse calibration information for a given spectrometer and probe combination. An additional command, getprosol, must be used to load the correct pulse calibration information from the "prosol" (probe and solvent information) table.

This proceeds as follows:

rpar PROTON all   load parameter set PROTON including all subsets
getprosol update   acquisition   parameters   with   correct calibrations

Note that in order for this to work correctly the "probe" variable must be set correctly, using the command edhead, and the prosol table must be set up correctly. We show an example of the prosol table in Appendix 4.1.

At the authors' facility we choose to operate in a different fashion. We set up and save our own customized parameter sets on each spectrometer for each probe in common use. Our parameter sets contain all the relevant calibration information, so the user simply has to load the correct parameter set prior to acquiring data, as follows:

rpar h1.cryo all   load local parameter set for ^{1}H observation using cryoprobe

The specifics of the standard parameter sets will vary somewhat for individual spectrometers and probes. Below are examples from our facility for a Bruker Avance DRX-500 spectrometer equipped with a 5-mm TXI cryoprobe.

^{1}H standard parameters (parameter set h1.cryo)

PULPROG:	zg30
TD:	32K (K = 1024)
NS:	8
DS:	2
SWH:	6510.40 (Hz, ca. 13 ppm)
AQ:	2.52 s
RG	set using rga or manually optimized for each sample
D1:	1.00 s
NUC1:	1H
P1:	8.0 μs
PL1:	(+)1.6 dB
SFO1:	499.89250 MHz
O1 (O1P):	2500 Hz (ca. 5.0 ppm)
SI:	16k
LB:	0.3 (Hz)

^{13}C standard parameters (parameter set c13.cryo)

PULPROG:	zgpg30
TD:	32k
NS:	256
DS:	8
SWH:	30303.03 (Hz, ca. 240 ppm)
AQ:	0.54 s
RG	usually the maximum value (or close to it) for your spectrometer
D1:	2.00 s
NUC1:	13C
P1:	16.0 μs
PL1:	−3 dB
SFO1:	125.71124 MHz
O1 (O1P):	13,800 Hz (ca. 110 ppm)
CPDPRG2:	waltz16
NUC2:	1H
PCPD2:	80.0 μs
PL2	120 dB
PL12:	21.0 dB
PL13:	24.0 dB
SFO2:	499.89200 MHz
O2 (O2P):	2000 Hz (ca. 4.0 ppm)
SI:	16k
LB:	3.0 (Hz)

APPENDIX 1.2 PROBE TUNING

If you have not carried out the probe tuning operation before, obtain the assistance of your spectrometer administrator or other qualified individual before attempting the following procedures.

Note: If your probe is equipped with motor-controlled tuning and matching please skip to the *Probe Tuning—Using ATMM or ATMA routine* section at the end of this Appendix.

Tuning the ^{1}H Coil—All Probes
1. We will tune the ^{1}H coil first. Turn off the sample spinning. Set up a ^{1}H dataset.
2. Type acqu to go to the acquisition window, then type wobb or cwobb (the latter for some systems with cryoprobes) to start the probe tuning ("wobble") routine.
3. The computer screen will show a trace with a dip in it; hopefully the dip will be close to the correct frequency in the center of the screen. If there is no dip, click the WOBB-SW button to the left of the screen, and enter a new value in the correct field of the screen that pops up. The default SW is 4 MHz; try 12 or 16 MHz if it needs changing.
4. When you have confirmed that there is a dip present, go to the magnet and adjust the yellow-labeled tune and match screws (screws may have large or small heads—do not confuse them with the probe mounting screws which are usually brass-colored) under the probe to bring the minimum of the dip to the center of the screen and also as close to the bottom of the display as possible.
   Note: The location of the dip left to right is referred to as the "tune" and the depth of the dip (closeness to the baseline) is referred to as the "match." The tune screw will tend to move the dip left–right and the match screw will improve the depth of the dip. The display on the preamp housing also shows the quality of the probe tuning and can be used in conjunction with or instead of the computer screen display.
5. When the tuning is OK, type halt to stop the wobble routine.
6. Restart sample spinning if desired. Leave spinning off for 2D experiments (Figure A1.2-1).

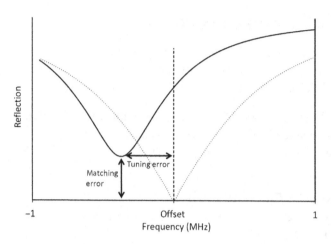

Figure A1.2-1 Probe tuning display (wobb command) of an arbitrary probehead. Dotted vertical line is the resonance frequency of the nucleus to be tuned. Solid red line: proper tuning has been achieved; solid black line: both tuning and matching are not correct.

Spectrometer Setup—X-Nucleus Tuning

1. Carry out steps 1−6 above to ensure the ^{1}H coil is tuned.
2. Use the `edc` or `re` command to create a new or read an existing dataset.
3. Use the `rpar` command to read the correct parameter set for the X-nucleus you want to observe, then type `ii` to initialize the spectrometer.
4. Type `acqu` to go to the acquisition window, then type `wobb` or `cwobb` (the latter for some systems with cryoprobes) to start the probe tuning ("wobble") routine.

Tuning the X-nucleus coil—broadband-type probes (BBO, BBFO, BBI, TBI, etc.)

5. X-nucleus coil tuning is done using gold-colored sliders visible on the bottom of the probe. The sliders are numbered, and there is a directory of numbers for the tune and match settings for most common X-nuclei hanging below the probe. Confirm that the tune and match settings are correct for the X-nucleus you want to observe.
6. The computer screen will show a trace with a dip in it; hopefully the dip will be close to the correct frequency in the center of the screen. If there is no dip, click the WOBB-SW button to the left of the screen, then enter a new value in the correct field of the screen that pops up. The default SW is 4 MHz; try 12 or 16 MHz if it needs changing.

7. When you have confirmed that there is a dip present, go to the magnet and adjust the gold tune and match sliders under the probe to bring the minimum of the dip to the center of the screen and also as close to the bottom of the display as possible.
   Note: The location of the dip left to right is referred to as the "tune" and the depth of the dip (closeness to the baseline) is referred to as the "match." The tune slider(s) will tend to move the dip left–right and the match slider(s) will improve the depth of the dip. The display on the preamp housing also shows the quality of the probe tuning and can be used in conjunction with, or instead of, the computer screen display.
8. When the tuning is OK, type halt to stop the wobble routine.
9. Be sure to tune the X-nucleus coil back to C13 before you leave the spectrometer.

Tuning the ^{13}C (and ^{15}N, Chapter 6) coils—,TXI or QXI-type probes

5. Proceed as above for H1 tuning, using the blue tune and match screws for tuning ^{13}C and the red tune and match screws for tuning ^{15}N (check first that these colors apply for your probe).
6. When the tuning is OK, type halt to stop the wobble routine.
7. Restart sample spinning if desired.
   Note: if your spectrometer has a QNP probe, consult with your spectrometer administrator prior to attempting any tuning of the X-nucleus coil.

Probe Tuning—Using ATMM or ATMA Routine

The ATMM/ATMA routine uses remote motor control to tune the probe so the operator does not have to physically manipulate the probe tuning controls at the probe. ATMM allows the operator to manually control the tuning and matching operation with a software interface, while ATMA carries out the process automatically.

1. After the desired experiment is set up, type atmm to start the probe tuning routine. If you are planning on running several experiments on the same sample, for probe tuning be sure to first set up the experiment which uses all the different nuclei which you will be measuring.

2. A wobb window will appear along with an atmm interface window. Click the arrow buttons in the TUNE and MATCH fields in the atmm window to optimize the probe tuning and matching as displayed in the wobb window.
3. Multinuclear experiments will result in a button being displayed in the atmm window for each active nucleus. Tune all relevant nuclei going from lowest frequency to highest. Click each nucleus button in turn to tune for that nucleus.
4. When all tuning is complete exit atmm using the File tab pulldown.
5. If your NMR facility or specific spectrometer uses only automated probe tuning, the command atma will start the automated process. You will still get a glimpse of the tuning curve(s) during the operation, similar to what you see when using atmm.

APPENDIX 1.3 QUININE SPECTRA

See Figures A1.3-1 and A1.3-2.

APPENDIX 1.4 LINESHAPE EFFECTS OF NONSPINNING SHIMS

See Figure A1.4.

APPENDIX 1.5 PULSE WIDTH CALIBRATION WITH PAROPT

The au program paropt provides the automation of the parameter optimization process. To carry out the p90 pulse width determination, proceed as follows. We assume here that a normal ^{1}H acquisition has just been completed as in Part 1, steps 1−3.

1. Confirm the current dataset is set up for observe-pulse calibration or adjust as necessary, typically with $NS = 1$, $DS = 0$, $D1 =$ a long delay (ca. 12 s for ^{1}H), $PL1 =$ the normal value for your spectrometer and phase correction is set. Reduce the current RG value by ca. 50%. Make sure the pulse sequence in use ($PULPROG$) does not include the text "30" in its title; e.g., use $PULPROG = $ zg, **not** zg30. Note that whatever plot region is defined now ($F1P$ and $F2P$ in edp) will be used when paropt displays the spectra, so alter it if necessary.
2. The following example will be approximately correct for an initial survey ^{1}H p90 calibration on most spectrometers.

Quinine in CDCl3, 298K, 500 MHz cryoprobe (#02).

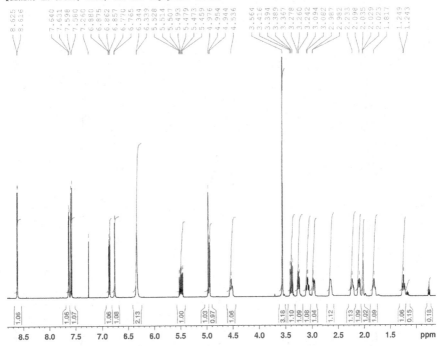

Figure A1.3-1 500 MHz ^1H spectrum of quinine hydrochloride in CDCl$_3$ at 298 K.

Quinine in CDCl3, 298K, 13C, 500 MHz cryoprobe (#02).

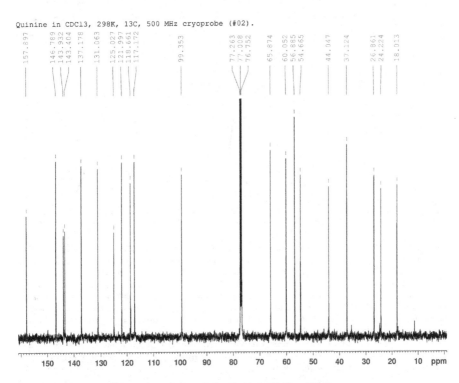

Figure A1.3-2 125 MHz ^{13}C spectrum of quinine hydrochloride in CDCl$_3$ at 298 K.

(a)

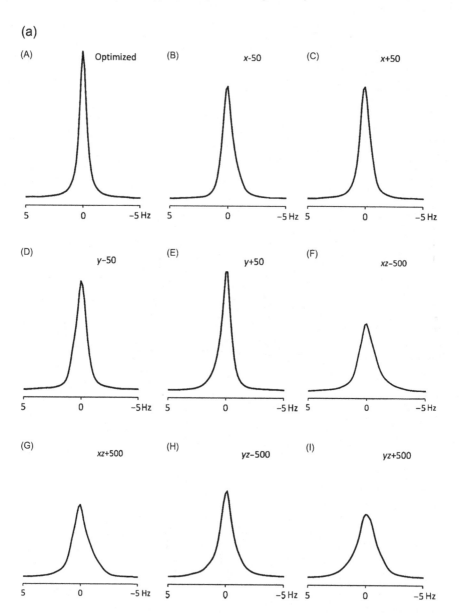

Figure A1.4 (a) Impact of various nonspinning shims upon the observed lineshape of the residual solvent signal from a CDCl$_3$ sample obtained using a Bruker Avance 500 MHz spectrometer equipped with a BBO probe and no sample spinning. The various shim conditions are: optimized (A); X shim is reduced (B) or increased (C) by 50; Y shim is reduced (D) or increased (E) by 50; XZ shim is reduced (F) or increased (G) by 500; YZ shim is reduced (H) or increased (I) by 500; XY shim is reduced (J) or increased (K) by 1000; X2 − Y2 shim is reduced (L) or increased (M) by 1000. (b) Impact of various nonspinning shims upon the observed lineshape and spinning sidebands of the residual solvent signal from a CDCl$_3$ sample obtained on a Bruker Avance 500 MHz spectrometer equipped with a BBO probe, with 20 Hz sample spinning. For clarity and compared to Figure A.4(a), the vertical display is increased by 10 times. The shim conditions are: optimized (A) with a small spinning sideband (marked by *); X shim is increased by 500 (B); XZ shim is increased by 2000 (C); XY shim is increased by 10,000 (D); X2 − Y2 shim is reduced by 10,000 (E).

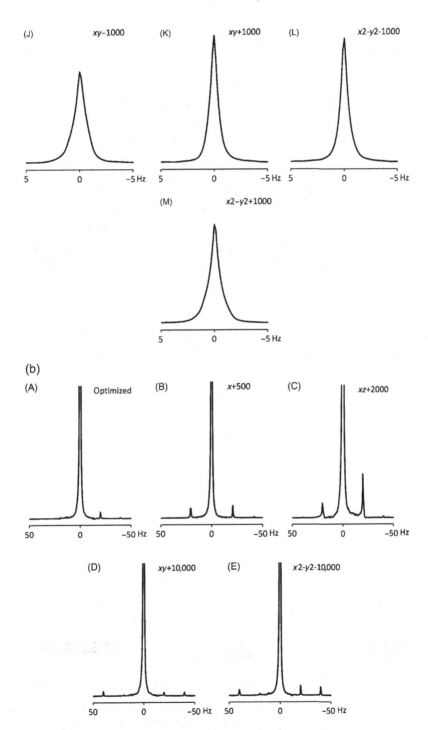

Figure A1.4 (Continued)

Type `paropt` to start the program. In sequence four windows will popup requesting input:

"Enter parameter to modify:"   p1 then click "ok"

"Enter initial parameter value:"   5 (μs) then click "ok"

"Enter parameter increment:"   5 (μs) then click "ok"

"Enter # of experiments:"   12 then click "ok"

3. The spectrometer now acquires 12 spectra with *P1* values of 5−60 μs in steps of 5 μs.

4. When finished all 12 spectra will be displayed end-to-end in procno 999 of the current experiment, along with a message window indicating `paropt` is finished and which *P1* value resulted in the maximum signal intensity (Figure A.5). Keep in mind that we are looking for the *P1* corresponding to the second null in the spectra display, which is the 360° pulse, and not the maximum intensity.

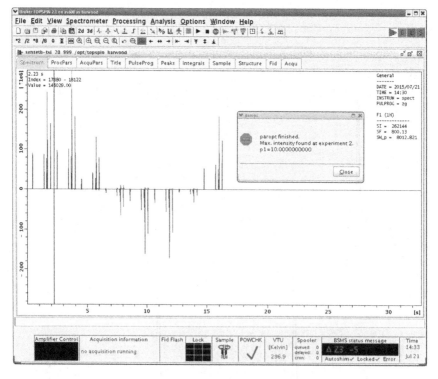

Figure A1.5-1 Display of a completed `paropt` *run obtained using the ^{1}H spectrum of the 0.1% ETB sample at 800 MHz. This shows p1 varying from 5 μs to 40 μs in steps of 5 μs (8 spectra). Note the cursor line denoting the approximate break between the first spectrum displayed and the second one adjacent. From this display we can discern that the p360 is somewhere between 35 and 40 μs, and further that it is much closer to 35 μs than to 40. Further investigation confirmed a p90 of 8.9 μs.*

5. When finished with the `paropt` display type `rep 1` to return to procno 1 of the current dataset.
6. Repeat as necessary with appropriately adjusted initial and incremental parameter values to home in on the precise p360. Always return to procno 1 before restarting `paropt`.

There is another program, `popt`, which does parameter optimization using a parameter input window with additional options and the ability to optimize multiple parameters. If you are familiar with this it may be used instead of `paropt`. If you want to try using `popt` for the p360 determination, set the processing parameter *PH_mod* to "pk" prior to starting `popt`.

CHAPTER 2

Multiple Irradiation and Multiple Pulse Experiments

OVERVIEW

This Chapter's exercises are designed to introduce you to some more-sophisticated NMR experiments. We will run several multiple irradiation and multiple pulse experiments, including a ^1H T_1 inversion-recovery (T_1-IR) measurement, ^1H homonuclear decoupling and selective nOe, ^{13}C acquisitions with and without nOe, and ^{13}C DEPT. Topics covered will include pulse sequences, homo- and heteronuclear decoupling, and T_1 data acquisition and analysis. We will use the quinine sample for these experiments, and the spectra obtained will assist in the chemical shift assignments for this molecule. The material covered in this chapter is in Chapters 2–4 of the Claridge book.

SAMPLE AND SPECTROMETER REQUIREMENTS

This chapter's exercises will use sample A. Since the NMR experiments in this chapter are simple 1D acquisitions, there are no special requirements for the spectrometer hardware. The combination of the sample concentration and the spectrometer's ^{13}C sensitivity should be such that the exercises in parts 4 and 5 can be completed in a reasonable amount of time.

ACTIVITIES

Part 1—^1H T_1 (Spin–Lattice) Relaxation Measurement

1. Put the quinine/$CDCl_3$ sample in the magnet. Lock and shim the sample and tune the probe for ^1H.
2. Run a ^1H spectrum using the standard parameters. Optimize the observe window for ^1H (OI and SW) and rerun the spectrum (we will also carry out this operation for 2D NMR experiments).

Practical NMR Spectroscopy Laboratory Guide: Using Bruker Spectrometers.
DOI: http://dx.doi.org/10.1016/B978-0-12-800689-4.00002-1

3. Copy the optimized spectrum to a new expno (`wrpa` / `re` or `new`). There are several steps we need to go through to set up the T_1-IR experiment. First, in `eda` set the pulse program (*PULPROG*) to "t1ir," then set the parameter mode (1,2,3 icon in the upper left) to 2D. To display the pulse sequence use `edcpul` or the PulseProg tab.

4. Now we need to edit the VD list file. The VD is the delay between the 180° inversion pulse and the 90° read pulse in the sequence. Type `edlist vd`, enter a filename in the "New file" field (e.g., "quin-t1.xyz" where xyz = your initials), and an editing window will open with a blank text file. Enter the following VD values as shown (one entry per line), without any units (default unit is seconds):

   0.001
   0.2
   0.4
   0.7
   1.0
   2.0
   4.0
   7.0

   Click the OK button to save the file and exit the editor.

   Note: For TopSpin versions 2, the procedure is a little different, as follows. Upon typing `edlist vd`, a "Parameter Lists" window will pop up. Go to the "File" pulldown in the upper left corner and select "New." Enter your filename in the "New Name" field of the second pop-up window, click the "OK" button, then enter the text in the new file window. When you are finished editing the file, go to the "File" pulldown, select "Close" then confirm that you want to save the changes. If the "Parameter Lists" window stays visible go to "File" → "Close" to remove it.

5. Go back to the AcquPars tab and enter the following:

FnMODE	QF
TD (F2)	16K
TD (F1)	8 (this is equal to the number of delays in the VD list)
NS	8
DS	4

6. Now type `ased` to enter the parsed acquisition parameter input (the software reads the pulse program and asks for the relevant pulse program inputs) and enter the following:

D1	12 (seconds) (relaxation delay)
VDLIST	Enter filename used in step 4 above
P1	Enter the value you obtained in Chapter 1 ^{1}H p90 calibration
PL1	Enter the value used in Chapter 1 ^{1}H p90 calibration

7. Now use the ProcPars tab or type `edp` and enter the following processing parameters:

SI (F2)	16K
SI (F1)	8
WDW (F2)	EM
LB (F2)	1
PH_mod (F2)	pk
PH_mod (F1)	no
MC2	QF

8. Start the experiment with `zg`. This will take ca. 25 min.
9. After the experiment has completed, type `rser 1` to read the first FID (of 8). In the new window, do `ef` and then phase this spectrum with all peaks inverted (this spectrum has the shortest VD). When exiting the phase routine use the save 2D icon or type `.s2d` which will save the phase corrections to the T_1-IR (2D) dataset. *Note*: You also could read the last FID (#8) and phase its spectrum positive (normal fashion) and save these corrections.
10. Type `to2d` to return to the original 2D dataset.
11. Process the data using `xf2` (Fourier transform in F2—the acquisition dimension). This will give a 2D display of the set of 8 spectra in a contour plot format.
12. Right-click on the 2D display to open "Display Properties," check "Show projections" and click "F2 projection only."
13. Right-click on the upper projection and select "External Projection" and in the menu that appears enter the expno of the 1D spectrum you ran prior to this T_1-IR run.
14. If you click the "oblique mode" button on the icon bar, or type `.st` you will see a 3D stacked plot. This is shown in Figure 2.1.

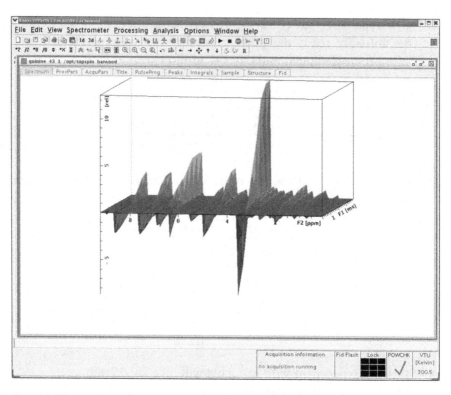

Figure 2.1 Oblique display of the quinine T_1-IR dataset after Fourier transformation in F2 (command xf2).

15. Type slice then select "Rows & Columns" then click "Interactive row/column display" and "OK," then click the "Scan rows" (↕) button and move the mouse pointer to look at the 1D spectra that make up the dataset. Click the "return" icon (↵) in the icon bar to exit this mode.

16. To calculate the T_1 values, go to the "Analysis" pulldown and select "T_1/T_2 relaxation" (see Figure 2.2 for an illustration of the relaxation calculation window):

a. Click "Extract Slice" from the NMR Relaxation Guide (right-hand side) pane, then in the pop-up window select "Spectrum" then enter 8 for the "Slice Number=" (to pick the last spectrum of the dataset) followed by "OK"

b. Click "Define Ranges" (rhs pane), then click the range icon (⌣, the left-most icon) in the icon bar above the displayed spectrum (like selecting a region in the integration mode); for TopSpin versions 2, click "Peaks/Ranges (rhs pane), then in the pop-up window select "Manual Integration" followed by "OK"

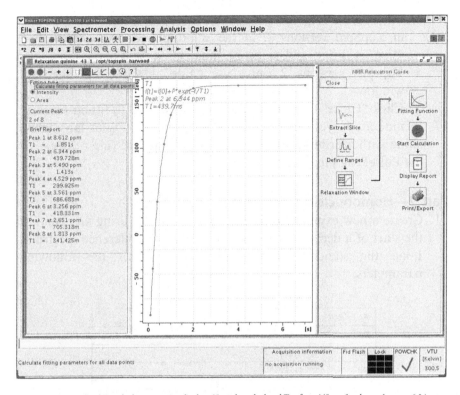

Figure 2.2 Example of T_1 calculation routine display. Note the calculated T_1 of ca. 440 ms for the peak at ca. 6.34 ppm.

   c. Select ranges (by left-clicking and dragging) 6–8 peaks from the spectrum for the T_1 calculation (we recommend only selecting single, well-separated peaks)

   d. Click the Save Region icon (the diskette picture) and then select "Export Regions to Relaxation Module & .ret."

   e. Click "Relaxation Window" (rhs pane), the defaults should be OK (in TopSpin versions 2 a "Relaxation Parameters" window pops up, click "OK" to keep the defaults), then choose the "Intensity" button in the "Fitting Type" pane which shows up on the left-hand side of the main window

   f. Click the "Calculate Fitting Parameters for All" button ($>>$) at the far left of the upper icon bar—this does the T_1 calculations for the peaks in the selected ranges

   g. Click "Display Report" (rhs pane), then in the pop-up text window click "Save as" (for Topspin versions 2, use the "File" pulldown → "Save as" then click the "Home" icon—looks like a little house) and enter a filename (e.g., quin-t1-calc.xyz, xyz as in step 4)

h. Click "Print/Export" (rhs pane) to obtain a hardcopy output; in TopSpin versions 2, in the still-displayed text window, use "File" → "Print" and confirm printing in the print pop-up windows

i. Topspin versions 2 only: use "File" → "Close" in the text window

j. Click "Close" (top of rhs pane) to exit the T_1/T_2 fitting routine

17. Include a copy of the T_1 calculations printout in your lab report, along with a printout of the T_1-IR pulse sequence (use `edcpul` or the PulseProg tab, then print the screen).

Part 2—Homonuclear Decoupling

1. Create a new expno or dataset—we recommend using an expno at the start of a decade (e.g., 51). This will be the reference spectrum. Load the standard parameters and then set the following parameters:

PULPROG	zg
D1	5 (seconds)
O1P	6 ppm
SW	14 ppm
P1/PL1	90° pulse (use your calibrated values from Chapter 1)
LB	0.5

Turn sample spinning off and recheck the shims. Check the *RG* then rerun this experiment. Use `abs` after phase correction.

2. Copy this experiment to an expno one unit higher (e.g., `wrpa 52` followed by `re 52`).

3. Set up the homodecoupling pulse sequence. First enter the pulse program "zghd.2" in the pulse program field of the AcquPars tab or the `eda` menu. Scroll down to the "Nucleus 2" field, click on the "Nucleus 2" link (TopSpin versions 2 only) then click "Edit." In the RF routing window that opens, set click the list field under "F2" and select ^1H, then click on the "Default" button at the bottom of the screen (TopSpin versions 2 only; this sets the RF routing. This can also be accomplished manually by clicking on the "F2" button followed by the "SGU2" button). Finally, click the "Save" button at the bottom of the window. Note that the RF routing window can be opened at any time using the `edasp` command.

4. Click the PulseProg tab to view the pulse program and the necessary parameters. Set the following parameters in the AcquPar tab or using `eda`:

AQ_mod	qsim or DQD (some spectrometers will require qsim for this parameter)
DIGMOD	Homodecoupling-digital
HDDUTY[%]	20

Then in `ased` set the following:

PL2	120 dB
PL24	63 dB

5. TopSpin uses the terms "O1" and "O2" to refer to the offset frequencies (from the spectrometer's base frequency) for the RF observe transmitter (the spectrum center frequency) and the RF decoupler, respectively. For homonuclear decoupling we will need to determine one or more values of *O2* to set up the decoupling properly.

6. Expand the spectrum about the peak you want to decouple. Click with the left mouse button on the Set RF icon in the icon bar above the spectrum or type `.seto123` . Then move the cursor to the desired decoupling frequency and left-click. In the O1/O2/O3 window that pops up, click on the O2 button with the left mouse button. This will set the *O2* value and exit the Set RF screen and return to the normal spectrum display.

7. Obtain the spectrum with decoupling and use `abs` for baseline correction. Examine the spectrum to see if the desired peak is completely decoupled. If necessary, increase the decoupling power by changing *PL24* from 63 to 60 dB, and repeat. Adjust again in −3 dB steps if necessary. Consult your spectrometer administrator you if you need to use a power level higher than (i.e., a *PL14* setting lower numerically than) 30 dB.
Note: If an error appears regarding "Invalid acquisition time..." try reducing the *SW* by ca. 1 ppm and retry the acquisition.

8. Using the multiple-spectrum display (click the icon or type `.md`, then type `re expno` where `expno` is the experiment number of the reference spectrum, e.g., 51 as given in step 1; alternatively use the browser window and the drag-and-drop function to load the reference spectrum data folder into the spectrum window), compare the current spectrum with the reference. Look for peak-shape changes due to the decoupling.

9. Copy this experiment to a new expno. Repeat the decoupling experiment using a different *O2* value (i.e., decouple a different peak), and check the decoupled spectrum as above.
10. Plot both decoupled spectra and the reference spectrum in a way that demonstrates the effect of the decoupling.

Part 3—nOe-difference

1. Copy your reference spectrum from Part 2 to a new decade (e.g., `re 51 wrpa 61` followed by `re 61`).
2. Copy your first decoupled experiment from Part 2 to the next expno (e.g., `re 52 wrpa 62` followed by `re 62`).
3. Set up the nOe-difference pulse sequence—enter the pulse program "zgf2pr" in the pulse program field of the AcquPars tab.
4. Click the PulseProg tab to view the pulse program and the necessary parameters. Set the parameters in the AcquPar tab or using `ased` according to the pulse program suggestions. Use 69 dB for the starting value of *PL14*. Be sure to reset the parameters *AQ_mod* to "DQD" and *DIGMOD* to "digital." These are the normal settings.
5. Expand the spectrum about the peak you want to irradiate and set the *O2* value as in Part 2.
6. Obtain the spectrum with irradiation, phase the spectrum carefully, and use `abs` after processing. Examine the spectrum to see if the irradiated peak is gone. If necessary, increase the irradiation power by changing *PL14* from 63 to 60 dB, and repeat. Adjust again in 3 dB steps if necessary.
   Note: The *PL14* power level selection should include the considerations that it be sufficient to saturate the intended signal effectively, but also have minimal impact upon adjacent signals. Further, *PL14* should not be higher than what the spectrometer or the sample can tolerate (given its long irradiation time). Consult your spectrometer administrator you if you need to use a power level higher than (i.e., a *PL14* setting lower numerically than) 30 dB.
7. Using the multiple-spectrum display (click the icon or type `.md`), compare the current spectrum with the reference. If the two spectra are on top of each other, you will see the missing, saturated, peak. Look closely for peaks which show a (slightly) higher intensity in the irradiated spectrum when compared to the reference—these peaks are from the positive nOe interaction which should be the case for the quinine sample.

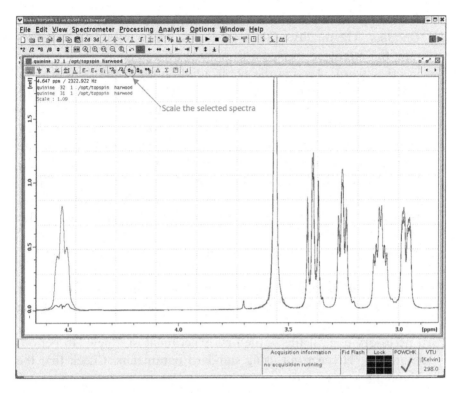

Figure 2.3 Multiple display mode display of an irradiated spectrum (at 4.56 ppm) and the reference spectrum on the quinine sample at 500 MHz, upfield expansion. Note the scale factor of 1.09 required to equalize the intensity of the peak at ca. 3.1 ppm in the irradiated spectrum relative to the reference spectrum, indicating an approximate 9% nOe.

8. To obtain an approximate value for the nOe enhancement in the multiple display mode, ensure that the two spectra are separated in the display using the "Toggle the display layout" icon (two spectra and arrow ↕). Then, click the unfilled box on the upper right-hand side of the display to select the reference spectrum for manipulation. Click the "Toggle the display layout" icon again to place both spectra on top of each other. Then, click the "Scale the selected spectra" icon (circled in purple in Figure 2.3) and adjust the vertical scale of the reference spectrum so that any peak of interest matches its intensity in the irradiated spectrum. You will see the scale factor on the left side of the display, usually this will be slightly larger than 1, e.g., 1.09, representing an approximate 9% nOe (Figure 2.3).

9. Copy this experiment to a new expno. Repeat the experiment using a different *O2* value (i.e., irradiate a different peak), check the irradiated spectrum as above.

10. Repeat for other *O2* values as desired with each in a new experiment.
11. *Optional*: The quality of the experiment can be improved by rerunning the reference spectrum to include off-resonance irradiation—in other words, leaving the *O2* irradiation on in the same way as for the nOe experiment but with the *O2* value set such that the irradiation is not on a peak. To do this, copy your first nOe experiment from step 3, above, to a new expno. Use the .set0123 command to set the *O2* to an area of the spectrum where there are no peaks, such as ca. 6 ppm. Set the *PL14* value to be equal to the highest power level you used for any of the peaks you irradiated. Now run this experiment, and after obtaining the spectrum repeat the multiple display exercise. See if your nOe percentages differ from the previous values.
12. Plot at least two irradiated spectra along with the reference spectrum.

Part 4—^{1}H-decoupled ^{13}C Spectra with and without nOe

1. Create a new dataset. Check or carry out locking and shimming with spinning on.
2. If necessary (e.g., if this is the first experiment of a new lab period) obtain the ^{1}H spectrum using standard parameters. Check that the shimming is acceptable.
3. New expno. Set up a ^{13}C experiment using standard parameters and $NS = 128$ or higher, if needed, to get a good signal. Check probe tuning on ^{1}H and ^{13}C. For reference we will call this expno 2.
4. Copy the ^{13}C experiment to expno 3. Change the pulse program from "zgpg30" to "zgig30." Use edpul or the PulseProg tab to see the differences between the two pulse sequences.
5. Go back to expno 2. If your spectrometer is running TopSpin 1.x, type multizg and enter 2 when asked for the number of experiments. This au program will then run expnos 2 and 3 in sequence. *Note:* the expnos must be sequentially numbered for multizg to work. This will take about 15 min if 128 scans are used for each expno. If your spectrometer is running TopSpin 2.x, you may use the software's queuing functionality instead. In this case, simply type zg in each expno you want to run and the software will run them one after the other. For queuing to work the expnos do not have to be sequential.
6. After multizg or all queued experiments complete, go back to expno 2 and process the data. Then process the data in expno 3. You also may process the FIDs directly after each is acquired.

7. In expno 2, touch up the phasing if necessary and apply `abs`. Pick a couple of peaks to check the signal-to-noise ratio of, using `sinocal`. Make a note of the peaks' locations, the noise region that you use, and the s/n values you obtain.
8. Repeat step 7 in expno 3 for the same peaks and noise region. Note the effect of the nOe irradiation on the s/n.

Part 5—^{13}C DEPT Spectra

1. Copy the ^{13}C expno 2 (step 3) to a new expno (e.g., 4). Set up a DEPT-45 experiment using `ased` and the following parameters:

PULPROG	dept
P1	^{13}C 90° pulse length (use your calibrated value from Chapter 1 labs)
PL1	Power level for your ^{13}C 90° pulse calibration
P3	^{1}H 90° pulse length (use your calibrated value from Chapter 1 labs)
PL2	Power level for your ^{1}H 90° pulse calibration
P0	45° ^{1}H pulse length
CNST2	145 (approximate one-bond J_{C-H} value in Hz)
NS	64 or as necessary to get a good signal (should be a multiple of 4)

2. Duplicate this expno to two new expnos such that all three are numerically in sequence. In the next expno, change the *P0* parameter to be a 90° ^{1}H pulse, and in the last expno set *P0* to be a 135° pulse (i.e., 1.5 * the 90° ^{1}H pulse length). This ^{1}H pulse duration determines the intensity and phase of the different types of ^{13}C peak as shown in the figure (Figure 2.4).
3. Go back to the first expno of the three and run all three experiments using either `multizg` (TopSpin 1.x) or the queuing functionality (TopSpin 2.x and higher). This will take about 20 min if each experiment has 64 scans.
4. After the data is acquired process the first spectrum (the DEPT-45) of the three and phase all peaks positive. Then use `multiefp` to process all three spectra with the same phase corrections. If your spectrometer does not have the au program `multiefp` installed, you must make a note of the phase correction parameters *PHC0* and *PHC1* obtained after phasing the DEPT-45 spectrum. Enter these phasing parameters into the DEPT-90 and DEPT-135 datasets before processing the datasets in turn using `efp`.
5. Use multiple display to show all spectra together and print that screen. This is shown in the figure below. Determine the number

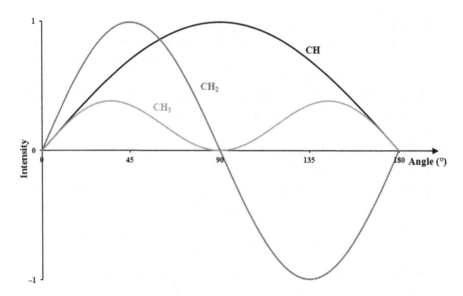

Figure 2.4 Expected intensity of −CH−, −CH₂−, and −CH₃ signals versus ¹H pulse P0 angle in the DEPT experiment.

and type of each ^{13}C resonance (e.g., $-CH-$, $-CH_2-$, $-CH_3$ and quaternary) (Figure 2.5).

6. Duplicate the DEPT-135 dataset to a new expno. We will now check the impact of changing the J_{C-H} value. Set up two new DEPT-135 datasets as done in step 2, above, and in one new dataset set $CNST2 = 130$ (Hz) and in the other set $CNST2 = 160$.

7. Run these experiments as in step 3 and process as in step 4. Compare the phasing behavior of the peaks, and create a multi-spectrum plot using multiple display to show this.

SUMMARY

Your lab report should include the following:

1. T_1 calculation printout and the T_1-IR pulse sequence printout.
2. Decoupling: Reference and at least two decoupled spectra using the same plot region.
3. NOe-difference: Plotted as in question 2 above.
4. ^{13}C spectra: Reference and no-nOe spectra using the same region with at least one peak's s/n values in the title.

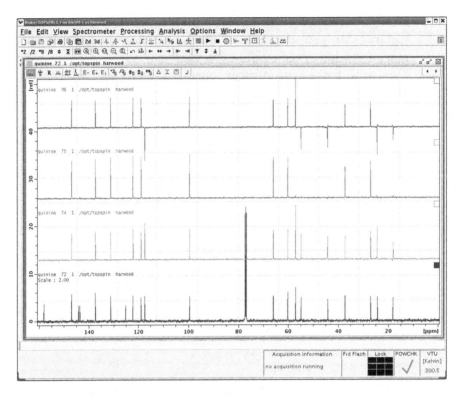

Figure 2.5 ^{13}C and DEPT spectra of quinine in CDCl$_3$ shown using the multiple display mode. From bottom to top: standard ^{13}C, DEPT-45, DEPT-90, DEPT-135. Note that the quaternary carbon resonances are not present in any of the DEPT spectra.

5. DEPT spectra: Screen print (as in Figure 2.5) of the reference and three DEPT spectra with different *P0* values, expanded as appropriate to get the best display.

6. DEPT spectra: Another screen print of the three DEPT-135 spectra with different J$_{C-H}$ values, expanded as appropriate to get the best display.

7. Suggested questions for consideration:

   a. What was the difference in signal-to-noise ratio for the selected peaks in the ^{13}C spectra obtained with and without nOe?

   b. Referring to question (a), is this s/n difference consistent with expectations? Why?

   c. Make up a table of each quinine ^{13}C chemical shift and its multiplicity (quaternary, −CH−, etc.)

d. How did changing the J-value optimization affect the DEPT spectra? Can you tell what the best J_{C-H} value setting is for quinine, and what criteria would you use?

e. Can you make any further chemical shift assignments for quinine?

CHAPTER 3

Polarization Transfer and Its Applications

OVERVIEW

This Chapter's work is designed to provide an introduction to one of the most important processes in modern NMR—polarization transfer. First we will use a sample of $CHCl_3$ to investigate the polarization transfer phenomenon with selective irradiation. From there we will generalize the process using the INEPT pulse sequence, followed by a look at its variants. Then we will use this example to provide an introduction to 2D NMR by way of the ^{13}C-detected HETCOR experiment. Finally, we will run two ^{13}C-detected heteronuclear correlation experiments using the quinine sample. These ultimately will be compared and with the ^{1}H-detected counterparts in Chapter 5. The material covered in this chapter is in Chapters 4 and 6 of the Claridge book.

SAMPLE AND SPECTROMETER REQUIREMENTS

This chapter's exercises will use samples E and A. Since the NMR experiments in this chapter are relatively undemanding, there are no special requirements for the spectrometer hardware. The combination of the sample A concentration and the spectrometer's ^{13}C sensitivity should be such that the exercises in Part 5 are able to be completed in a reasonable amount of time.

ACTIVITIES

Part 1—Polarization Transfer via Selective Irradiation
1. Put the $CHCl_3/CDCl_3$ sample E in the magnet and lock, shim, and tune the probe for ^{1}H.
2. Run a ^{1}H spectrum using the standard parameters. Print out the entire spectrum.
3. Optimize the observe window for ^{1}H ($O1$ and SW), using the icon or .setsw, around the $CHCl_3$ line and rerun the spectrum.

Practical NMR Spectroscopy Laboratory Guide: Using Bruker Spectrometers.
DOI: http://dx.doi.org/10.1016/B978-0-12-800689-4.00003-3

Use appropriate values of *TD* and *SI* for a ca. 2-s acquisition time. Make a note of your *O1* value and the *SW* value.

4. You should see the $CHCl_3$ signal and the two ^{13}C satellite peaks. Measure and record the $^1J_{C-H}$ value.

5. Using the command .set0123, determine the decoupler (*O2*) values required to center ^1H irradiation on each of the two satellite peaks. Record these two *O2* values.

6. Using a new expno, set up a ^{13}C acquisition starting with the standard parameters, and then set *NS* = 1, *DS* = 0, *D1* = 6 s. Make a note of the default value for *O2*—this is the default center frequency for the ^1H CPD decoupling irradiation for routine ^{13}C acquisitions. Check the probe tuning for ^{13}C, then acquire the data.

7. Expand the ^{13}C spectrum from ca. 85−65 ppm and use the icon or .setsw to set the observe window to this range. Use *TD* = 4K and *SI* = 2K to get a reasonable acquisition time (ca. 1 s). Reacquire the data.

8. Copy the data to a new expno. Set *PULPROG* to "zg" and *D1* = 12 s, reacquire. This will produce a ^1H-coupled ^{13}C spectrum.

9. Check and record the signal-to-noise ratio of each of the $CHCl_3$ ^{13}C doublet peaks using sinocal. Use the same noise region (e.g., 2 ppm from 65 to 75 ppm) for this and every subsequent sinocal measurement (in this lab).

10. Copy the data to a new expno. We will now acquire the first FID with ^1H CW selective polarization. We will need to modify an existing Bruker pulse program to do this. Type edpul and select the pulse program "zggd." Once the pulse program is open in the editor, replace the line "4u cpd2:f2" with "4u cw:f2." This changes the decoupling from using a CPD (broadband-decoupling) sequence to using a single-frequency (selective) irradiation. Save the pulse program using a new filename such as "zggd.xyz" (xyz = your initials). It is good practice to add a comment line (beginning with a semicolon (;)) at the start of the pulse sequence file describing any changes made to the file.

    Note: You may need to save your new pulse program before editing, if so, use the "File" → "Save as" pulldown to do so, using the new filename as given above.

11. In the new expno set *PULPROG* to be the new pulse program. Set *PL12* (**do not** accidentally change *PL2*!) to a very low power level, such as 69 dB. Set *O2* to match one of the *O2* values you recorded in step 5 for the $CHCl_3$ ^{13}C satellites in the ^1H spectrum.

12. Acquire the data and process. Get the signal-to-noise ratio for the positive peak of the ^{13}C doublet using the same settings as in step 9.
13. Copy to a new expno. Set the *O2* to the other satellite peak, rerun and check the s/n ratio of the positive peak.
14. Try adjusting the value of *PL12* by ±3 or 6 dB, rerun, and see if that affects the s/n ratio.

Part 2—Polarization Transfer via INEPT

1. Now we will use the INEPT pulse sequence (Claridge figure 4.23a) to do the polarization transfer. Copy the previous dataset to a new expno, then set the following in `ased`:

PULPROG	ineptnd
CNST2	$^1J_{C-H}$ value
P3, PL2	^{1}H 90° pulse values (as measured in Chapter 1, Part 3)

Finally, **set O2 to the default value from the first ^{13}C acquisition** in step 6—note that we are now not irradiating any specific ^{1}H signal! Take a look at the INEPT pulse program using the PulseProg tab and the graphical display window.

2. Acquire the data, process, and phase the spectrum so that the larger doublet component is positive. Measure the s/n ratio of the positive peak as before.
3. Duplicate this data to a new expno. We will now refocus the doublet components so that both are in phase. To do this, set the following:

PULPROG	ineptrd (rd = refocused, decoupled; Claridge figure 4.23b)
CNST11	4 (refocusing time—optimized for CH)
PL12	120 dB (turn off decoupling during data acquisition)

Acquire and process with both peaks positive.

4. Duplicate this data to a new expno. We will now decouple the refocused doublet by setting up CPD decoupling using the standard ^{1}H$-^{13}$C decoupling parameters, as follows:

CPDPRG2	waltz16
PCPD2	Use the value from your spectrometer's standard ^{13}C parameters
PL12	Use the value from your spectrometer's standard ^{13}C parameters

Rerun and process. Confirm that you now obtain mainly a singlet ^{13}C resonance.

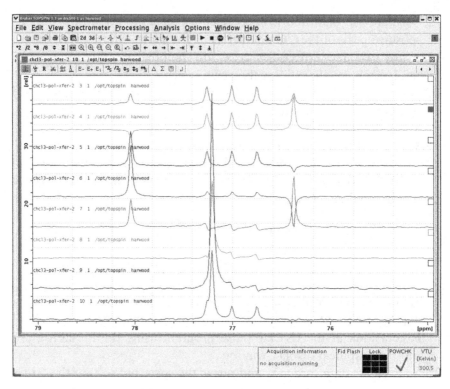

Figure 3.1 *Various* ^{13}C *spectra of* $CHCl_3$—*sequence (top to bottom) follows the text order in which the spectra were acquired, starting from Part 1, Step 8.*

5. Duplicate this data to a new expno. Set $D1 = 4$ s and $NS = 4$. Rerun. This will reduce the $CDCl_3$ signal relative to the $CHCl_3$ signal by phase-cycling. Measure the s/n ratio of the now singlet peak using the same noise region as in step 6.

6. Duplicate this data to a new expno. We will now run a normal ^{13}C acquisition for comparison. Set $PULPROG = $ zgpg, rerun, and check the s/n ratio again (Figure 3.1).

Part 3—Heteronuclear Correlation Experiment Setup and Acquisition

1. Copy your last INEPT dataset to a new expno. We are now going to set up a 2D heteronuclear correlation data acquisition from scratch. Keep in mind that F2 is the acquisition dimension which

for this experiment is ^{13}C and F1 is the indirect dimension which for this experiment is ^{1}H. In eda set the following:

PULPROG	hxcoqf (^{13}C–^{1}H correlation, magnitude mode, ^{13}C-detected—i.e., ^{13}C = F2)
1-2-3	1→2 (change 1D–2D)
FnMODE	QF
TD	2K (F2), 64 (F1) (number of data points acquired)
O2P	Same as O1P for ^{1}H spectrum in Part 1, step 3 (^{1}H center frequency)
ND_010	2 (if present)
SW	Same as ^{13}C (F2), same as ^{1}H spectrum in Part 1, step 3 (F1)

Note: **ND_010 (if present) must be set before the SW (F1) value is entered!**

2. Now use ased and enter the following:

NS	4
DS	0
D1	3 (seconds)
CNST2	210 (^{1}J$_{C-H}$ for CHCl$_3$)
CNST11	3 (optimize sequence for all C–H types)
P1/PL1	90° pulse for ^{13}C (use your calibrated value from Chapter 1 labs)
P3/PL2	90° pulse for ^{1}H (use your calibrated value from Chapter 1 labs)
CPDPRG2	waltz16 (CPD program for ^{1}H decoupling)
PCPD2/PL12	^{1}H p90 calibration values for decoupling (use the standard values for your spectrometer, as above)

3. Now go into edp or the ProcPars tab and set the following processing parameters:

SI	2K (F2), 128 (F1) (number of data points in the 2D spectrum)
WDW	EM (F2), SINE (F1) (window function in each dimension)
LB	5 (line broadening used in F2 with EM)
SSB	1 (F1, sine-bell function)
PH_mod	PK (F2), MC (F1) (phasing mode for each dimension)
MC2	QF (same as FnMode in the acquisition parameters menu)

4. Turn sample spinning off and recheck the shims, then run the experiment. Check probe tuning for ^{1}H and ^{13}C if it has not yet been done. The data acquisition will take about 15 min.

5. Compare the hxcoqf and ineptrd pulse sequences. Note that they are very similar once the 180° refocusing pulses present in the ineptrd sequence have been removed.

Part 4—Heteronuclear Correlation Experiment Data Processing

1. First we will take a look at this dataset to demonstrate in a very basic way how the 2D acquisition works. Type xf2 to carry out a Fourier transformation in the F2 (^{13}C) domain. This will result in a set of 64 ^{13}C spectra.

2. Type slice to enter the interactive 2D display mode. Select the "interactive" button in the pop-up menu (or use the command .md instead of slice).

3. Move the crosshair to display the interferogram (data in F1 dimension) of highest intensity. Note the oscillation frequency of this data. This shows the chemical shift dependency of the magnetization transfer from the ^{1}H nucleus to the ^{13}C. This is how the 2D data is generated—note how this "interferogram" mimics a FID. The oscillation frequency of this interferogram determines the chemical shift of the peak in F1 after Fourier transformation in the F1 (vertical) dimension. This generates a cross-peak at the CHCl$_3$ ^{1}H frequency (Figure 3.2).

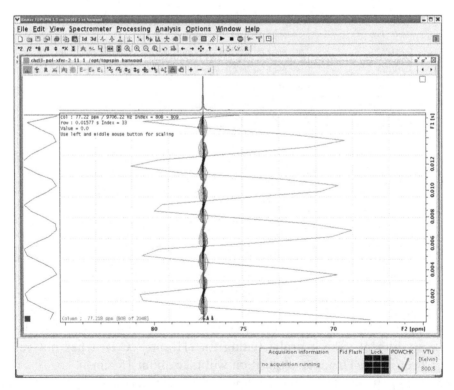

Figure 3.2 2D Heteronuclear correlation spectrum of CHCl$_3$ after Fourier transformation in the F2 (^{13}C) domain. Note that the interferogram taken from the most-intense data is displayed as the vertical projection.

4. Now type `xf1` to carry out the Fourier transformation in F1 and complete the 2D FT. You should see a contour plot with a peak at the CHCl$_3$ chemical shift in both the ^{13}C and ^{1}H dimensions. Try to adjust the vertical scale (*2 or /2) if no contours appear. Note that the command `xfb` carries out `xf2` followed by `xf1`.

5. This 2D experiment is decoupled in both dimensions, i.e., the ^{1}H$-^{13}$C correlation will appear as a singlet.

6. On the top of the 2D dataset a projection will be displayed. To show the original ^{13}C spectrum here, right-click in the projection field and select "External" for the projection type, then fill in the correct experiment number for the ^{13}C spectrum in the menu that pops up.

7. On the left of the 2D dataset a projection will be displayed. To show the original ^{1}H spectrum here, right-click in the projection field and select "External" for the projection type, then fill in the correct experiment number for the ^{1}H spectrum in the menu that pops up.

8. Adjust the vertical scale to get a good representation of the vertical intensity. Then right-click on the contour plot and select "Set Contour Levels." In the menu that pops up, adjust the number and spacing of the levels if desired, then click "Fill" and "Apply" to set the new levels and to save the vertical scaling setting. Usually we recommend to reduce the "Level increment" parameter in the contour level menu from the default value of 1.8 to a value in the range of 1.5−1.2. Further, we usually increase the "Number of levels" from 8 to 14−20. These changes will result in a better-resolved contour display and plot (this will be important for 2D experiments we perform later in the book).

9. Print the 2D data with both projections displayed (Figure 3.3).

Part 5—Heteronuclear Correlation Experiments on Quinine—Experiment Setup

1. Replace sample E with the quinine sample A. In a new expno, obtain the ^{1}H spectrum using the previously-used parameters for quinine. Optimize the observe window (don't include the 12 ppm peak) and rerun. Make a note of the *O1* and *SW* values you used for the optimized spectrum.

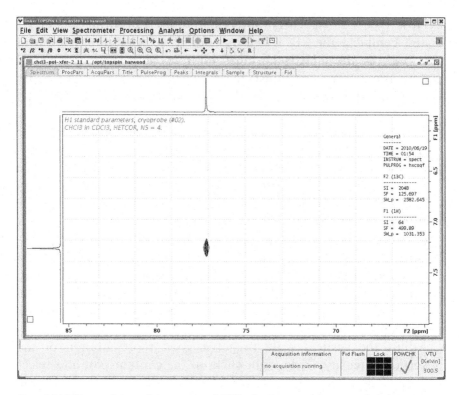

Figure 3.3 2D Heteronuclear correlation spectrum of CHCl₃ after Fourier transformation in both dimensions.

2. In a new expno, set up ^{13}C acquisition. Tune the probe for both nuclei. Run the ^{13}C then optimize the observe window to include all the ^{13}C peaks, and rerun.

   Note: Strictly speaking, the ^{13}C observation window for HETCOR does not need to include quaternary carbons. Therefore, if the ^{13}C spectrum of your sample has one or more quaternary carbon resonances as its most downfield peaks, as is often the case, these may be excluded when setting up the optimized ^{13}C acquisition prior to running a HETCOR experiment. In the current example there is little to be gained from eliminating the most downfield resonance in the ^{13}C spectrum so for simplicity we use the entire spectrum. This also removes the need to use a different ^{13}C observation window for the COLOC experiment we will run later.

3. Copy the optimized ^{13}C spectrum to a new expno. Set up the HETCOR experiment as described in steps 1, 2 and 3 of Part 3, above. Use the same pulse sequence. For the ^{1}H (F1) dimension use the *O1* and *SW* values from the optimized ^{1}H spectrum you obtained in Part 5, step 1 (above), and set the following in eda and ased.

TD	4K (F2), 128 (F1) (number of data points acquired)
ND_010	2 (if present)
SW	Same as ^{13}C (F2), same as ^{1}H spectrum in Part 5, step 1 (F1)
O2P	Same as *O1P* for ^{1}H spectrum in Part 5, step 1 (^{1}H center frequency)

Note: **ND_010 (if present) must be set before the SW (F1) value is entered!**

4. Now use ased and enter the following:

NS	8 (or more if needed, use a multiple of 4)
DS	8
CNST2	145 ($^1J_{C-H}$ average value)

5. Now go into edp or the ProcPars tab and set the following processing parameters:

SI	4K (F2), 512 (F1) (number of data points in the 2D spectrum)
PH_mod	PK (F2), MC (F1) (phasing mode for each dimension)
WDW	EM (F2), SINE (F1) (window function in each dimension)
LB	8 (line broadening used in F2 with EM)
SSB	1 (F1, sine-bell function)

6. Copy the HETCOR experiment to a new expno one digit higher than the current one. We will use this dataset as a reference to setup the COLOC experiment.

7. The COLOC (COrrelation by LOng-range Couplings) experiment is very similar to HETCOR but it is designed to show correlations due to long-range (2- and 3-bond) C−H couplings. We will optimize this experiment for a long-range coupling constant of 6 Hz. In this particular example we will use the same sweep widths (in both dimensions) for the COLOC experiment as we

used for the HETCOR experiment. Note that the COLOC experiment must be set up to include all ^{13}C resonances in F2, so often it will use a larger observation window (i.e. ^{13}C sweep width) than the HETCOR spectrum. See also the NOTE in step 2, above.

8. In the new expno, set up the following using ased:

PULPROG	colocqf
D6	42 ms (1/2 J for J$_{lr}$ = 6 Hz)
D18	28 ms (1/3 J for J$_{lr}$ = 6 Hz)
NS	16 (or more if needed, use a multiple of 8)

Note: If you are running this experiment on a 300 MHz (or lower frequency) spectrometer you will need to reduce the *TD* parameter in F1 (*TD1*) from 128 to 96 increments.

9. Go back to the HETCOR expno. Turn sample spinning off and recheck the shims, then run both experiments by either using multizg (TopSpin 1.x) or the queuing functionality (TopSpin 2.x and higher). The data acquisition will take about 2 h (with *NS* values of 8 and 16) but each acquisition may be stopped early if necessary, at the cost of resolution in the FI (^{1}H) dimension of the 2D plots.

Part 6—Heteronuclear Correlation Experiments on Quinine—Data Processing

1. Go to the HETCOR dataset and use xfb to process.
2. Go to the COLOC dataset and use xfb to process. Both these experiments are magnitude mode experiments so there is no phasing needed.
3. After processing set up the projections in each dimension as described in Part 4 steps 6 and 7.
4. Locate the CHCl$_3$ peak in each 2D spectrum (or some other known peak), expand it, and set its chemical shift in both dimensions to the correct value (icon or command) prior to plotting.
5. Print out both the HETCOR and COLOC spectra with projections—include projections in both dimensions (Figure 3.4).

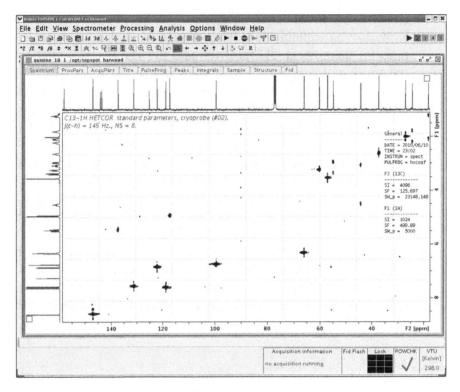

Figure 3.4 2D Heteronuclear correlation spectrum of quinine at 125/500 MHz.

SUMMARY

Your lab report should include the following:

1. ^1H spectrum of the $CHCl_3$ sample using the full observation window. Ensure the vertical scaling (cy) is set high enough to show the ^{13}C satellite peaks of the $CHCl_3$ peak.
2. All the relevant ^{13}C spectra of $CHCl_3$ (single-pulse, selective polarization and INEPT) with s/n ratios in the title text.
3. HETCOR spectrum of $CHCl_3$—with correct projections.
4. HETCOR and COLOC spectra of quinine—with appropriate projections.

5. Suggested questions for consideration:
   a. How do you explain the difference in measured signal-to-noise ratio for the normal ^{13}C spectrum and the selective polarization spectra (Part 1, step 9 vs. Part 1, steps 12 and 13)?
   b. How do you explain the difference in measured signal-to-noise ratio for the INEPT spectrum compared to the selective polarization spectra (Part 2, step 2 vs. Part 1, steps 12 and 13)?
   c. How do you explain the difference in the measured signal-to-noise ratio for the decoupled ^{13}C spectrum compared to the decoupled INEPT spectrum (Part 2, step 6 vs. step 5)?
   d. What do you think would have happened if we had tried to obtain a decoupled INEPT spectrum of $CHCl_3$ without setting up the correct refocusing delay (Part 2, steps 3 and 4)?
   e. Can you make any further chemical shift assignments for quinine using the heteronuclear correlation spectra?
   f. Is the HETCOR spectrum of quinine consistent with the DEPT spectra, in particular with respect to the $-CH_2-$ resonances? Explain.

CHAPTER 4

Homonuclear Correlation Experiments

OVERVIEW

This Chapter's work is designed to familiarize you with several homonuclear 2D correlation experiments. We will be obtaining ^{1}H COSY (COrrelation SpectroscopY), TOCSY (TOtal Correlation SpectroscopY), and NOESY (Nuclear Overhauser Enhancement SpectroscopY) spectra on the quinine sample. We will set these experiments up "from scratch" starting with the 1D ^{1}H spectrum, without using any standard parameter sets. We will also go over the processing of the 2D data, including setting the chemical shift reference, window functions, and 2D phasing—all of these experiments, other than the first one, are phase sensitive. These spectra will provide useful information towards the assignment of the quinine ^{1}H spectrum. The material covered in this chapter is in Chapters 5 and 8 of the Claridge book.

SAMPLE AND SPECTROMETER REQUIREMENTS

This chapter's exercises will use sample A. Since the NMR experiments in this chapter are relatively undemanding, there are no special requirements for the spectrometer hardware (we do not use gradient-enhanced pulse sequences for this chapter's work).

ACTIVITIES

Part 1—Quinine ^{1}H Spectrum

1. Put in the quinine sample A and lock, shim, and tune the probe for ^{1}H. In a new expno (e.g., 101), run the ^{1}H spectrum, then optimize the observe window (this time, do include the broad 12 ppm peak if it is present in your sample) and rerun. Make a note of the $O1$ and SW values you used for the optimized spectrum (typically ca. 5.5 ppm and 13 ppm). Make sure to set the chemical shift reference.

Practical NMR Spectroscopy Laboratory Guide: Using Bruker Spectrometers.
DOI: http://dx.doi.org/10.1016/B978-0-12-800689-4.00004-5

Part 2—COSY Setup

1. Copy the optimized ^1H spectrum to a new expno one digit higher than the current one. We will now set up the homonuclear COSY data acquisition from scratch. In this experiment both the acquisition dimension (F2) and the indirect dimension (F1) are ^1H. In eda set the following:

PULPROG	cosyqf ($^1H-^1H$ correlation, magnitude mode)
1–2–3	1→2 (change 1D–2D)
FnMODE	QF (magnitude mode)
TD	2K (F2), 256 (F1) (number of data points acquired)
ND_010	1 (if present)
SW (Hz)	F2 and F1 same as ^1H spectrum in Part 1, step 1

Note: **ND_010 (if present) must be set before the SW values are entered!**

2. Now use ased and enter the following:

NS	4
DS	16
RG	Same as 1D (Part 1, step 1)
D1	2 (seconds)
P0/P1/PL1	90° pulse for ^1H (use your calibrated value from Chapter 1, Part 3, step 7)

3. Now use edp or the ProcPars tab and set the following processing parameters:

SI	2K (F2), 1K (F1) (number of data points in the 2D spectrum)
WDW	SINE (F2), SINE (F1) (window function in each dimension)
SSB	1 (both dimensions)
PH_mod	No (F2), MC (F1) (phasing mode for each dimension)
MC2	QF

Part 3—TOCSY Setup

1. Copy the COSY experiment to a new expno one digit higher than the current one. We will now set up the TOCSY data acquisition using the COSY experiment as a template. In eda set the following:

PULPROG	mlevph ($^1H-^1H$ correlation via spinlocking, phase sensitive)
FnMODE	States-TPPI (phase sensitive)

2. Now use `ased` and enter the following:

NS	4
RG	Same as 1D (Part 1, step 1)
D9	80 ms (TOCSY mixing time)
P17	2.0–2.5 ms (trim pulse)
P6/PL10	90° pulse for ^{1}H spinlock

Note 1: If an error window appears with a message about reading parameters and *P7* and calculations, etc., after typing `ased`, execute the command `getprosol` prior to attempting `ased` again. This error occurs if values of 0 are present for some pulse lengths which are used in internal calculations carried out by the pulse sequence. Loading default pulse values with `getprosol` should remedy this. Alternatively, enter a value of ca. 30 μs for pulse *P6*.

Note 2: *P6* should be approximately 30 μs to 24 μs duration for 500–800 MHz spectrometers. This will require less RF power than what is used for the normal 90-degree pulse, which corresponds to a more-positive value of *PL10* when compared to *PL1*. You may calibrate this pulse manually using the method outlined in Chapter 1, Part 3, step 3, or you may use `edprosol` to see the software-determined values. Please see Appendix 4.1 for a short description of using `edprosol`.

3. Now go into `edp` or the ProcPars tab and set the following processing parameters:

SI	2K (F2), 1K (F1) (number of data points in the 2D spectrum)
WDW	QSINE (F2), QSINE (F1) (window function in each dimension)
SSB	3 (F2), 3 (F1) (shift for QSINE—you may experiment with different values)
PHC0	(F1) 180 (Phase corrections in F1 dimension)
PHC1	(F1) -180
PH_mod	pk (F2), pk (F1) (phasing mode for each dimension)
FCOR	(F1) 1
MC2	States-TPPI

Note: The values for PHC0, PHC1, and FCOR are given in the pulse program text file.

Part 4—NOESY Setup

1. Copy the TOCSY experiment to a new expno one digit higher than the current one. We will now set up the NOESY data acquisition using the TOCSY experiment as a template. In `eda` set the following:

PULPROG	noesyph (^{1}H–^{1}H correlation via nOe interaction, phase sensitive)

2. Now use `ased` and enter the following:

NS	8
RG	Same as 1D (Part 1, step 1)
D1	4 (seconds)
D8	300 ms (NOESY mixing time)

3. Now go into `edp` or the ProcPars tab and set or check the following processing parameters:

SI	2K (F2), 1K (F1) (number of data points in the 2D spectrum)
WDW	QSINE (F2), QSINE (F1) (window function in each dimension)
SSB	3 (F2), 3 (F1) (shift for QSINE function)
PHC0	(F1) 90
PHC1	(F1) -180
PH_mod	pk (F2), pk (F1) (phasing mode for each dimension)
FCOR	(F1) 1
MC2	States-TPPI

Part 5—2D Data Acquisition

1. Go back to the COSY dataset from Part 2 (using `edc`, `re`, etc.).
2. Turn off sample spinning and recheck the shims.
3. Run the experiments sequentially by either using `multizg` (TopSpin 1.x; confirm that the experiments are in consecutive expnos) or the queuing functionality (TopSpin 2.x and higher). If scheduling requirements preclude acquiring all three datasets sequentially (ca. 4 h) in one session we would suggest acquiring the NOESY data (ca. 2.5 h) during a separate session.

Part 6—COSY Data processing

1. Go to the COSY dataset.
2. Do the 2D Fourier transform using `xfb`.

   Note: You may experiment with using different window functions and/or processing parameter settings for data processing—simply change the parameter of interest in `edp` or the ProcPars tab and execute `xfb` again to see the result of the change.
3. This COSY experiment is a magnitude-mode experiment and does not need to be phased.
4. Right-click in each of the projection areas and load the original ^{1}H spectrum into each.
5. Expand the solvent peak on the diagonal (or another peak with a known chemical shift) and set the chemical shift referencing.
6. Set the desired vertical intensity and then set the contour levels, cf., Chapter 3, Part 4, step 8.
7. Expand as desired (left-click and drag), set the plot region by right-clicking on the spectrum and selecting "Save Display Region To..." and then "Parameters F1/2..."
8. Plot the 2D spectrum with both projections. Plot expansions as necessary. Include a descriptive title.

Part 7—TOCSY Data Processing

1. Go to the TOCSY dataset.
2. The TOCSY is a phase-sensitive experiment and it should show all peaks positive. It must be phased manually. The easiest way to do this is to first determine the phase corrections in F2 by reading one of the FIDs and Fourier transforming it, phasing the resulting spectrum, and saving the phase corrections to the 2D dataset. Proceed as follows:

   a. Read the second FID of the dataset using `rser 2`.

   b. Type `lb` and enter 10, do `ef`, then phase the resulting spectrum with all peaks positive.

   c. When finished phasing, use `.s2d` or click the save-2D icon to save the phase corrections to the TOCSY dataset, followed by `.ret` or clicking the return icon.

   d. Type `to2d` to return to the 2D dataset.

3. Do the 2D Fourier transformation using `xfb`. Because of the *PHC0* and *PHC1* inputs made in `edp` for F1, the spectrum should be phased reasonably well in both dimensions. If the phasing needs improvement, do the following:
   a. Enter the phasing routine.
   b. Move the cross-hairs to a diagonal peak on the upfield corner of the spectrum.
   c. Right-click on this peak and select "Add" from the pop-up menu.
   d. Repeat steps (b) and (c) for both a peak in the center and in the downfield corner.
   e. Click the "R↔" icon in the phasing submenu to switch to the row display.
   f. You will see the three spectra displayed together with the Pivot Point displayed.
      Phase the spectra (all three will change together) so that all peaks are positive.
   g. Click the "Return & save" icon (or type `.sret`) to save the corrections.
   h. You will now see the phased 2D spectrum in the phasing subroutine display. If the phasing looks OK click the ↵ (return) icon or type `.ret`.
   i. If the phasing needs more work use the "R↔" or "C↕" icons and repeat the process as above on either the rows ("R") or columns ("C") as needed. If you do additional phase correction on the rows the columns will need further correction.
4. Right-click in each of the projection areas and load the original ^{1}H spectrum into each.
5. Expand the solvent peak (or another peak with a known chemical shift) and set the chemical shift referencing.
6. Set the desired vertical intensity and then set the contour levels. For the TOCSY data you may choose to plot only the positive contours for a cleaner appearance. Alternatively plot just one negative contour.
7. Expand as desired (left-click and drag), set the plot region by right-clicking on the spectrum and selecting "Save Display Region To…" and then "Parameters F1/2…"
8. Plot the 2D spectrum with both projections, and expansions as necessary (Figure 4.1).

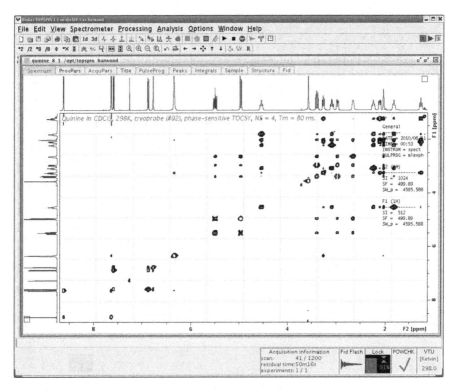

Figure 4.1 TOCSY spectrum of quinine obtained at 500 MHz, positive contour levels plotted.

Part 8—NOESY Data Processing

1. Go to the NOESY dataset.
2. The NOESY is also phase sensitive, but it should show the diagonal peaks as negative. For a small molecule, the nOe cross-peaks (off-diagonal) will be positive (opposite phase from the diagonal). Any cross-peaks originating from chemical exchange will be negative (same phase as the diagonal). To do the phasing in F2 carry out the same procedure as for the TOCSY data (Part 7, step 2), **EXCEPT use FID #1 instead of #2** (i.e., rser 1 in step 2a) and phase the spectrum to be negative (inverted).
3. Do the 2D Fourier transformation using xfb. If the spectrum needs additional phasing, carry out the same procedure as for the TOCSY data (Part 7, step 3), **EXCEPT in step (f)** phase the spectra so that the diagonal peaks in each trace are negative (inverted). Other peaks in each trace may end up being positive or negative (Figure 4.2).
4. Follow steps 4 through 8, above, as for TOCSY for the rest of the processing. Be sure to include **BOTH positive and negative** contour levels on the plot(s).

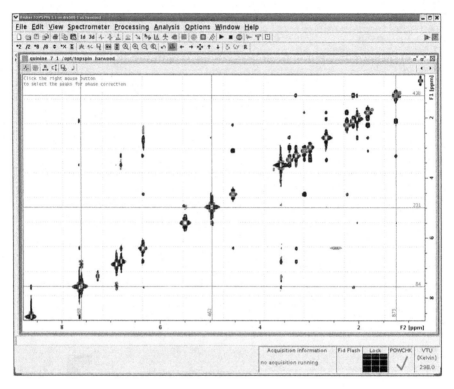

Figure 4.2 Top-level display of 2D phasing routine of the NOESY spectrum of quinine. Red contours are negative intensity. The figure shows three row/column slices selected with cross-hairs. The row icon R (↔) can be used to display the selected rows for phasing. These rows are shown in Figure 4.3.

SUMMARY

Your lab report should include the following:

1. Printouts of each 2D spectrum with projections and descriptive titles.
2. Suggested questions for consideration:
   a. Are there any cross-peaks present in the TOCSY experiment which were not present in the COSY? Are they reasonable, based on your assignment expectations?
   b. On the NOESY there may be cross-peaks present with both positive and negative phases. Can you explain the origins of these two types of cross-peaks?
   c. Please update/expand your list of chemical shift assignments for quinine.

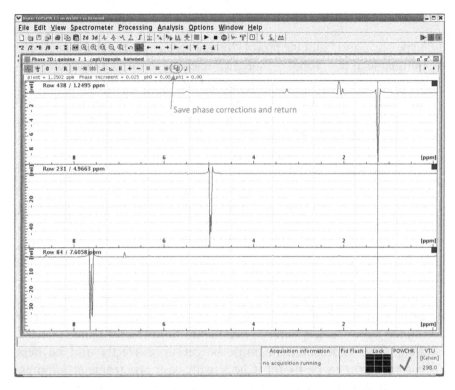

Figure 4.3 Row display of 2D phasing routine of the NOESY spectrum of quinine. The figure shows the three rows selected in Figure 4.2 with the reference peak for the first-order phase correction shown by the red line. The diagonal peak of each row is negative and the other peaks in each row are positive. Note that the icons in the upper left of the window have changed. The 0 and 1 icons are used to apply phase corrections (as in 1D phasing), and the "save and return" icon (circled in purple) will save the phase corrections and return to the upper-level display shown in Figure 4.2.

APPENDIX 4.1 USING edprosol TO DETERMINE THE POWER LEVEL FOR THE TOCSY SPINLOCK PULSE *P6*

The spinlock component of the TOCSY experiment uses a lower-power pulse (*P6*) than the normal 90° pulse calibration, such that the 90° pulse used for the spinlock ranges from about 30 μs for a 500 MHz spectrometer to about 24 μs for an 800 MHz spectrometer. You could calibrate this pulse manually, as done in Chapter 1, or you could use the edprosol feature of TopSpin to determine the correct power level setting. This command accesses a database which holds all the default pulse calibrations for commonly-used pulses for a given probe. It can also give you calibration values for various pulses based on the nominal p90 pulse calibration. The following assumes that the edprosol database is correct and kept up-to-date.

Proceed as follows:

1. Type edprosol and the prosol ("prosol" is an assumed (by the authors) portmanteau of "probe" + "solvent," referring to the fact that pulse calibration parameters are generally dependent on these two factors) window will open up. Check that the text describing the probe in the "Probe's name" field is consistent with the currently-installed probe. If you have questions about this check with your spectrometer administrator.

2. Select ^{1}H from the nucleus window and the default ^{1}H calibrations for hard pulses will be displayed. Check that the entries in the "pulse [μs]" and "power level" fields for the hard 90° pulse ("P90") are consistent with the values you obtained in Chapter 1, if not, enter the correct values.

3. Go to the line for the "tocsy spinlock" pulse in the "F1" column, check that the pulse duration is the desired value, highlight the power level field, then click the "Calc" button. This will then give you the calculated power level for that pulse duration. Make a note of this, then exit edprosol and do not save changes (you will not be able to save changes without administrator access) (Figure A4.1.1).

   Note: The edprosol table must be set up correctly and be kept current for it to be useful. Check with your spectrometer administrator to confirm that this is the case for your spectrometer.

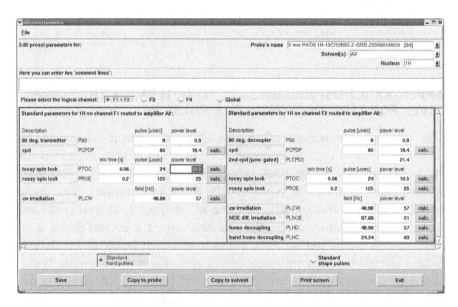

Figure A4.1.1 Example of edprosol display from TopSpin 2.1. The power level field for the TOCSY pulse P6 is highlighted in blue.

CHAPTER 5

PFG (Pulsed Field Gradient) Experiments

OVERVIEW

This chapter's exercises will introduce you to the pulsed field gradient (pfg) capability of the modern NMR spectrometer. We will first demonstrate the field gradient and then calibrate the field-gradient strength using the 1D image of a water sample. Then we will run gradient versions of some of the 2D experiments we have already run, so that you can see how the gradient can both speed up data acquisition and reduce artifacts in the 2D spectra. The first 2Ds we run will be gradient COSY, NOESY, and DQF-COSY (DQF = double-quantum-filtered). Then we will investigate the inverse-mode for detecting heteronuclear $^{13}C-^{1}H$ correlations, and we will run the pfg-HMQC and -HMBC experiments. These are the newer analogs of the HETCOR and COLOC experiments (respectively), which we ran in Chapter 3. The HMQC and HMBC experiments work by observing the ^{1}H signal, instead of the ^{13}C signal measured by the older experiments. This gives the new experiments a huge advantage in sensitivity, but it requires pfg capabilities to eliminate unwanted artifacts. This is because the ^{1}H signal we desire is from only the 1.1% of the ^{1}H resonances which are coupled to ^{13}C. The material covered in this chapter is in Chapters 3 and 6 of the Claridge book.

SAMPLE AND SPECTROMETER REQUIREMENTS

This chapter's exercises will use samples F and A. Since we are investigating pulsed field gradients the spectrometer and probe must be equipped with the appropriate field gradient hardware.

Practical NMR Spectroscopy Laboratory Guide: Using Bruker Spectrometers.
DOI: http://dx.doi.org/10.1016/B978-0-12-800689-4.00005-7

ACTIVITIES

Part 1—1D Z-Gradient Image of an H₂O/D₂O "Phantom"

1. We will use a "phantom" sample for the first part of this lab—this is a type of sample used in NMR imaging which has water (usually) in a specific spatial arrangement. This is employed to calibrate the field gradients used in imaging, and we do the same with our phantom sample today. For instructions on making the phantom please see Appendix 5.1.

2. Holding the phantom sample, measure the distance between the two white teflon vortex plugs. We will need to know this distance for the gradient strength calibration.

3. Put the phantom into a spinner and carefully center the liquid volume around the coil-center mark on the sample-spinner depth gauge.

4. Put the phantom sample into the magnet.

5. DO NOT SPIN the sample. Lock the sample on D_2O. Coarsely shim using Z1, Z2, and Z3.

6. Run the ^1H spectrum using the standard parameters. Center the transmitter frequency (O1) on the water peak and then set the water peak chemical shift to 0 ppm/Hz. Rerun.

7. Copy the spectrum to a new expno. We will now set up the 1D imaging experiment.

8. In the new dataset, use `eda` and set the following:

AQ_mod	qsim

9. Now open `ased` to enter the following:

PULPROG	imgegp1d (1D Z-axis gradient image)
TD	4K
NS	2
DS	0
SWH	100,000 (Hz)
D1	1 s
D15	5 ms
D21	250 μs
D27	5 ms
P0	3 μs
PL1	6 (dB) (we want to use only a small excitation pulse for this experiment)
GPZ1	−20 (%) (dephasing gradient—Z-axis)
GPZ2	10 (%) (acquisition gradient—this is the gradient strength being calibrated)

Note: If your spectrometer will allow a larger *SWH*, or requires a smaller *SWH* than given here, ensure that the signal is refocused during the acquisition by confirming that the absolute value of GPZ2*AQ is larger than that of GPZ1*D27. A good starting point is (GPZ1*D27) = (0.2)*(GPZ2*AQ).

10. Check the receiver gain (`rga` or manually) then run this acquisition using `zg`. You will see an echo signal in the FID display. Pick an appropriate window function to apply to the data (e.g., QSINE or GM) and adjust the window's parameters to match the echo center.

11. Apply your selected window function and Fourier transform the data, then do `mc` to get the magnitude-mode *Z*-axis image. Toggle the axis units to Hz (h/p icon).

12. Using the distance measurement tool, measure the width of the image and record the value. An example of this measurement is shown in Figure 5.1.

13. Double both gradient values and repeat the experiment (new expno). This run may require either higher pulse power, higher receiver gain,

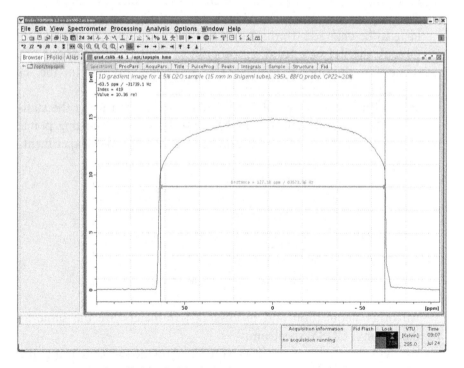

Figure 5.1 Measuring the width (in Hz) of the phantom's image.

or more scans to get a good signal-to-noise ratio. Measure the width of the image signal and record the value. This test will check the linearity of the Z-axis gradient strength.

14. Finally, look at the imgegp1d pulse sequence and print out a copy of the graphical display.

    Note: In the next chapter we will use the field gradient equipment to shim the samples in H_2O/D_2O that we will use.

Part 2—pfg-Homonuclear 2D Experiments Setup

1. Put in the quinine sample A and turn spinning on. Lock the sample, tune the probe, and shim the sample. *Optional*: If your spectrometer is capable of performing ^2H gradient shimming (TopSpin versions 1) or TopShim (TopSpin versions 2) it could be employed at this point instead of manual shimming. Check with your spectrometer administrator about the use of ^2H gradient shimming or TopShim (TopShim is a more-automated version of gradient shimming introduced in the later TopSpin release). There is further discussion of these topics starting at Chapter 6, Part 1, step 3. Note that the Chapter 6 discussion refers to ^1H gradient shimming; for ^2H gradient shimming the Shimming Method choice (step 4) will have to be changed from ^1D to $^1D^2H$.

2. Obtain the ^1H spectrum in procno 1 of your new experiment name. Use the same parameters (*O1* and *SW*) you used for the ^1H spectrum of quinine in Chapter 4, Part 1.

3. We will now set up the pfg-COSY data acquisition using the standard COSY experiment from Chapter 4, Part 2 as a starting point. Copy the COSY dataset from Chapter 4, Part 2 to a new experiment.

4. In the new experiment set the following in `ased`:

PULPROG	cosygpqf (magnitude-mode gradient COSY)
NS	1 (compare to 4 scans used for conventional COSY)
D16	200 µs (microseconds)
GPNAM1	SINE.100 (shaped pulse file for gradient pulse)
GPZ1	10 (gradient intensity in % of maximum)
P16	1 ms (milliseconds—gradient pulse duration)

All other acquisition and processing parameters are the same as the original COSY spectrum.

5. We will now set up the pfg-NOESY data acquisition using the standard NOESY experiment from week 4 as a starting point.

Copy the NOESY spectrum from Chapter 4, Part 4 to a new dataset with the same experiment name as the pfg-COSY above, but with the expno incremented by 1.

6. In the new experiment set the following in `ased`:

PULPROG	noesygpph (phase-sensitive gradient NOESY)
NS	2 (cf. 8 previously) (increase in steps of two scans if needed)
D16	200 µs
GPNAM1,2	SINE.100
GPZ1	40
GPZ2	−40
P16	1 ms

All other acquisition and processing parameters are the same as the original NOESY spectrum.

7. Repeat the same process to set up pfg-DQF-COSY using the NOESY experiment you just set up. Choose the correct target expno for the new dataset if `multizg` will be used.

8. In the new experiment set the following in `eda`:

PULPROG	cosydfetgp.2 (DQF-COSY gradient-selected, echo-antiecho)
FnMODE	Echo-Antiecho (This is a type of phase-sensitive 2D data acquisition specific to certain types of gradient-selected experiments)

9. In `ased` set the following:

NS	2 (this number may need to be increased in steps of two depending on your spectrometer's sensitivity)
D1	3 s
D13	5 µs (if requested)
D16	200 µs
GPNAM1,2,3	SINE.100
GPZ1	30
GPZ2	10
GPZ3	50
P16	1 ms

10. Be sure to set the processing parameter in `edp` *MC2* to "Echo-Antiecho" as well. Set *PHCO* (F1) to 90. All other parameters are the same as the NOESY spectrum.

11. *Optional*: Because the cross-peaks in the DQF-COSY spectrum show a fine structure, this experiment will benefit from acquiring the data with a higher resolution in F1 than we have used for the other homonuclear 2D experiments. Depending on the amount of time available to acquire the data, you may wish to increase the value of *TD* in F2, up to 512 points. If so, in eda set the following (**bolded** entries are the ones being changed):

TD	2K (F2), **512** (F1) (number of data points acquired) and then in edp or the ProcPars tab increase the *SI* parameter in F1 as well:
SI	2K (F2), **2K** (F1) (number of data points in the 2D spectrum)

Part 3—2D pfg-Homonuclear Experiments Data Acquisition

1. While still in the pfg-DQF-COSY dataset, type rga and allow the routine to finish. Take the value of *RG* obtained by rga and reduce it by ca. 50%. We need to check the *RG* for this experiment because the signal intensity will be much lower than for the others. However, the rga routine may underestimate the signal build-up that can occur as the acquisition progresses. Hence the reduction of the rga -obtained value. For the other experiments we will use the prior *RG* values.
2. Go back to the pfg-COSY dataset.
3. Turn off sample spinning, check shims.
4. Run the experiments sequentially by either using multizg (TopSpin 1.x) or the queuing functionality (TopSpin 2.x and higher). Check each experiment's approximate time requirement using expt in each experiment to estimate the total time required before you begin.

Part 4—pfg-Homonuclear 2D Experiments Data Processing

1. Go back to the pfg-COSY dataset.
2. For the data processing, follow the steps for the analogous non-gradient experiment as in Chapter 4, Parts 7 and 9. For the DQF-COSY spectrum, proceed as follows.
3. Go to the DQF-COSY dataset.
4. The DQF-COSY is phase-sensitive. In this case all the peaks (diagonal- and cross-) will show up as antiphase multiplets, which makes it difficult to phase this data without having had some practice. We suggest the following procedure to provide a reasonable starting point for the phasing in F2. Carry out the same steps as for the NOESY data (outlined in Chapter 4, Part 9, step 2). With

the current experiment, each peak will show dispersive (negative) components on its upfield and downfield edges after phasing the center of all the peaks positive. While this is not correct, strictly-speaking, it will provide a starting point for the final adjustment of the phasing in the next step.

5. Do the 2D Fourier transform using xfb. You should see a 2D spectrum which is phased close-to-correctly in both dimensions. To complete the phasing process, carry out the same procedure as for the NOESY data (Chapter 4, Part 9, step 3), **EXCEPT that in step (f)**, each peak in each of the selected traces will show both positive and negative intensity. Figure 5.2 shows part of the phasing process for this dataset, using as examples three F2 traces.

6. Follow Chapter 4, Part 9, steps 4 through 8, as done for NOESY for the rest of the processing. Be sure to include **BOTH positive and negative** contour levels on the plot(s) (Figure 5.3).

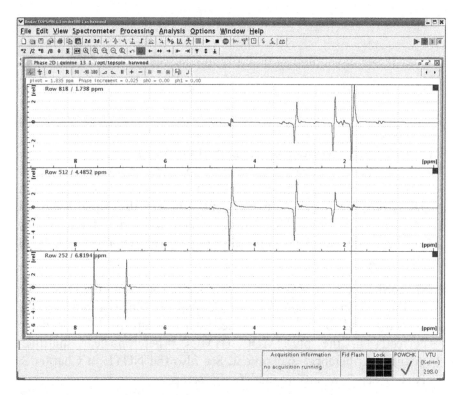

Figure 5.2 2D phasing routine for the pfg-DQF-COSY spectrum of quinine. Figure shows three F2 traces with peaks showing positive and negative intensity.

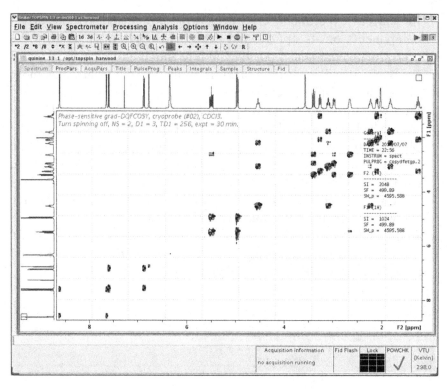

Figure 5.3 pfg-DQF-COSY spectrum of quinine at 500 MHz. Black contours are positive intensity, red ones are negative.

Part 5—pfg-Heteronuclear 2D Experiments Setup

Note: The heteronuclear experiments you are about to run are a testament to the importance of the pfg capability to a modern NMR spectrometer. These experiments observe the signal from ^1H nuclei correlated with ^{13}C nuclei—which are only 1.1% of the ^1H nuclei in the sample—while filtering out the remaining 98.9% of the ^1H signals.

1. Go to the ^{13}C experiment you ran prior to setting up the HETCOR experiment in Chapter 3, Part 5, step 2. Make a note of the *O1* value and the *SW*—this is needed to set up the ^{13}C dimension correctly in the pfg-HMQC (HMQC = heteronuclear multiple-quantum coherence) experiment. See also the NOTE in Chapter 3, Part 5, step 2, regarding the ^{13}C *SW* setup.

2. We will now set up the pfg-HMQC data acquisition using the standard COSY experiment from Chapter 4 as a starting point. Copy the COSY spectrum from Chapter 4, Part 2, to a new experiment. If you want to run this sequentially with the other experiments in this lab use the correct experiment name and number. In the new experiment set the following in `eda`:

PULPROG	hmqcgpqf (magnitude-mode gradient HMQC)
NUC2	13C, then "save," c.f. Chapter 2, Part 2, step 3 for the specific steps for your software version; this step sets up the decoupling (F2) channel properly
NUC1 (F1)	13C—this step is important!—sets the correct frequency reference in F1
O2P	*O1P* from ^{13}C spectrum
ND_010	2 (if present)
SW (F1)	*SW* from ^{13}C spectrum

Note: **ND_010 (if present) and NUC1 (F1) must be set before the SW (F1) value is entered!** The *NUC2* entry sets up the decoupling (i.e., the second RF) channel for the correct nucleus (in this case ^{13}C). The *NUC1(F1)* entry ensures that the software knows that the nucleus/frequency to be used for referencing the frequency calculations in F1 is ^{13}C. If this is not correct, the sweep width calculations will be wrong. Be aware, however, that these two parameters (*NUC2* and *NUC1(F1)*) are not always the same.

3. In `ased` set the following:

NS	2
RG	2k (we are only observing 1% of the ^{1}H spins) or check with `rga`
CNST2	145 (Hz, the one-bond J_{C-H})
D1	2
D16	200 μs
P1/PL1	^{1}H p90 calibration
CPDPRG2	garp (CPD sequence for broadband decoupling of ^{13}C)
PCPD2	50–70 ms (for 800–500 MHz spectrometers)
PL12	ca. 12 dB (power level for 50–70 μs ^{13}C p90—check using `edprosol`)
GPNAM1,2,3	SINE.100
GPZ1	50
GPZ2	30
GPZ3	40.1
P16	1 ms

4. In edp set the following:

SI	2k in both dimensions
SF (F1)	Set to the BF2 for ^{13}C
WDW	QSINE in both dimensions
SSB	0 in both dimensions
PH_mod	No (F2), mc (F1)

5. After setting all these parameters, duplicate this dataset to the next expno and set the following in ased:

PULPROG	hmbcgplpndqf (heteronuclear multiple bond correlation, with gradients, magnitude-mode)
CNST13	6 (Hz, long-range J_{C-H})

Note the instructions and comments for the COLOC experiment ^{13}C SW setup (Chapter 3, Part 5, step 7) apply here also.

Part 6—2D Data Acquisition

1. Go back to the pfg-HMQC dataset.
2. Check that sample spinning is off and that the shims are OK.
3. Run both experiments sequentially by either using multizg (TopSpin 1.x) or the queuing functionality (TopSpin 2.x and higher).

Part 7—pfg-Heteronuclear 2D Experiments Data Processing

1. Read the HMQC dataset and use xfb to process.
2. Read the HMBC dataset and use xfb to process. Both of these datasets are magnitude-mode experiments so they do not need phasing.
3. For both, set/check the projections, referencing and contour levels in each experiment after 2D Fourier transformation. You may use any previously-obtained ^{13}C spectrum of quinine for the external ^{13}C (F1) projection.
4. Compare the results of these spectra with the analogous spectra from Chapter 3 (Figure 5.4).

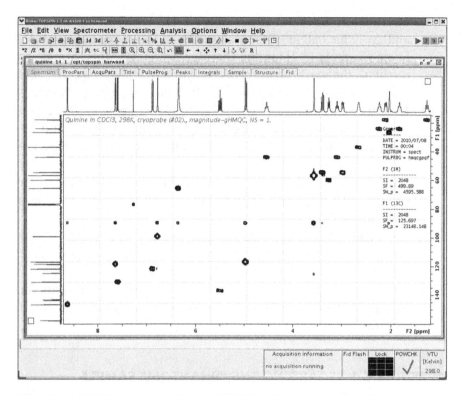

Figure 5.4 pfg-HMQC spectrum of quinine at 500/125 MHz. Compare to the HETCOR spectrum shown in Figure 3.4.

SUMMARY

Your lab report should include the following:

1. Printouts of the 1D image spectra at both gradient strengths.
2. If you carried out ^2H gradient shimming (or TopShim) of the quinine sample (optional), include a printout of the unshimmed quinine spectrum compared with the gradient-shimmed one.
3. Printouts of each 2D spectrum you obtained (there are five of them) with correct projections in both dimensions and descriptive titles. Check the chemical shift referencing in both dimensions.
4. Suggested questions for consideration:
   a. This formula is used to calculate the Z-axis gradient strength (G_z) in G (Gauss)/cm:

$$G_z = \Delta\nu/(4358 \times \Delta_z)$$

where $\Delta\nu$ is the width of the image in Hz as shown in Figure 5.1 and Δ_z is the height of the water column in the phantom sample in cm (formula (3.21) in the Claridge book). Use this formula to calculate the gradient strengths for your two image spectra, and report the results. Are the two results consistent with each other?

b. Describe the window function and its parameters which you used for the gradient-echo processing in Part 1, steps 10 and 11. Why did you choose this function and how did you pick the specific related parameters?

c. Can you comment on any differences between the homonuclear spectra you obtained in Chapter 4 compared to the ones obtained in this lab?

d. Can you make any comparisons between the heteronuclear experiments you obtained in this lab to the analogous ones obtained in Chapter 3 on the quinine sample?

e. Please update/expand your list of chemical shift assignments for quinine, if necessary (based on your current compilation of all spectra).

APPENDIX 5.1 REGARDING THE "PHANTOM" SAMPLE

The phantom sample can be made up using either a Shigemi™ tube set or a conventional 5 mm NMR tube and two 5 mm teflon vortex plugs. The Shigemi™ tube will give the better result, due to it providing a sharper transition between the water volume and the nonwater volume. In either case we recommend using a volume of liquid of about 1 cm height. Too large of a volume may cause distortions in the image if the sample height begins to approach the height of the probe's RF coil. Too small of a volume will reduce the signal intensity and make the measurement of the spectroscopic width of the resulting image more difficult.

When working with the vortex plugs, the first plug has to be inserted into the NMR tube with the threaded portion first, and then pushed down to the bottom of the tube. This has the unfortunate effect of making it impossible to remove this vortex plug from the NMR tube. However, we need to install the plug this way in order to have the flat side adjacent to the liquid volume. After adding the approximately correct amount of liquid plus a 10–20% excess, insert the second vortex plug in the normal fashion (threaded side up). Lower the plug through the liquid to get the desired height of the liquid column.

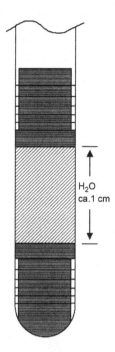

Figure A5.1.1 Diagram of the H_2O/D_2O phantom used in this exercise. Diagram depicts two teflon vortex plugs in a 5-mm NMR tube, separated by ca. 1 cm of H_2O/D_2O. The diagram attempts to depict the orientation of the plugs as described in the text, with the flat parts interfacing with the liquid volume.

Figure A5.1.1 illustrates the arrangement. For the liquid we recommend using H_2O in D_2O at ca. 20% (v/v).

If using a Shigemi™ tube set, simply add the correct amount of liquid plus a small excess to the outer tube and then insert the plunger piece to get the correct liquid height. Make sure to secure the plunger piece with either a dedicated cap or with Parafilm®, to keep it from moving relative to the outer tube.

Note: When making the sample, either with vortex plugs or with a Shigemi™ tube set, do your best to eliminate any air bubbles in the liquid volume, since they will reduce the quality of the images. The vortex plug typically has a tiny hole in its center to assist with this.

CHAPTER 6

Introduction to NMR of Biomolecules in H_2O

OVERVIEW

For this chapter we will turn our attention to a special sub-field of modern NMR spectroscopy, specifically, the spectroscopy of biomolecules in water. First, we will use a standard sample of 5 mM lysozyme in 90% H_2O/10% D_2O to investigate a variety of water-suppression techniques, and to run the frequently-used 2D NOESY experiment. Then, we will use another standard sample, this time 1 mM ^{13}C- and ^{15}N-labeled ubiquitin, also in 90% H_2O/10% D_2O, to acquire heteronuclear correlation experiments based on the HSQC scheme. For both samples we will investigate gradient shimming and pulse calibration. The material covered in this chapter is in Chapters 6 and 10 of the Claridge book.

SAMPLE AND SPECTROMETER REQUIREMENTS

This chapter's exercises will use samples G and H. If you do not have these two commercial samples available you may substitute your own protein or peptide samples. Keep in mind that any sample you use to replace the ubiquitin sample must be both ^{13}C- and ^{15}N-labeled if you want to run both of the HSQC experiments presented in this chapter. However, if you have a protein/peptide with just one label (e.g., ^{13}C or ^{15}N) you could run only the experiment involving that nucleus and skip the other.

A spectrometer equipped with at least three RF channels (^{1}H, X, and Y) and an inverse triple-resonance Z-gradient probe (typically ^{1}H–^{13}C–^{15}N or ^{1}H–^{13}C–BB with BB tuned to ^{15}N, usually referred to as a TXI or TBI probe) would be the ideal configuration to perform all the experiments in this chapter. If you have only a two-channel spectrometer at your disposal, however, the experiments will still work as we have presented them, since for simplicity's sake we do not employ

Practical NMR Spectroscopy Laboratory Guide: Using Bruker Spectrometers.
DOI: http://dx.doi.org/10.1016/B978-0-12-800689-4.00006-9

decoupling of the second X nucleus (Y nucleus) when acquiring the HSQC data. Since we are using pulsed field gradient experiments the spectrometer and probe must be equipped with the appropriate field gradient hardware. Also, in order for water suppression to work well while obtaining good sensitivity, the probe should be of the "inverse" type or at least be specified to perform water suppression. If this is not the case the experiments may still be carried out but the results will be sub-optimal.

ACTIVITIES

Part 1—Lysozyme Spectra—Shimming and Pulse Calibration

1. Put in the lysozyme sample, lock it (use the solvent "$H_2O + D_2O$"), allow the sample temperature time to equilibrate (ca. 5 min) then tune the probe. Do not shim the sample. Leave sample spinning OFF. Load the ^1H standard parameters and set the following in a sed:

TD	16K
NS	1
DS	0
SWH	16 (ppm)
RG	1
D1	2
P1	1

Then set the following processing parameters:

SI	16K
LB	1

2. Obtain the ^1H spectrum using the current shims. Copy this spectrum to procno 2 (wrp 2). Most likely you will only see a large peak from water in this spectrum; very high vertical expansions may reveal weak protein peaks. If the signal from the water peak is too intense (i.e., the spectrum is distorted even with $RG = 1$), reduce the *P1* duration until you are able to get an undistorted spectrum.

3. We will now use ^1H gradient shimming to shim the sample. The type of gradient shimming procedure we will use depends upon the version of TopSpin you are using. If you are using a spectrometer with TopSpin version 1.x, please skip to step 4. If you are using a

spectrometer with TopSpin version 2.x or newer, please keep reading this paragraph. Starting with TopSpin version 2 Bruker introduced a new automated shimming routine called "TopShim" which is based on gradient shimming. TopShim may be executed with no user input beyond entering the command topshim. It also may be invoked from the "Spectrometer"→"Shim control" pulldown/pull-right menu; this method actually starts the command topshim gui which brings up a menu requiring some user input, although the default settings usually suffice. Therefore, use either of these methods (topshim command or menu-driven) to start TopShim, and wait for it to complete. After it has completed, skip to step 6. *Note*: Gradient shimming as described in step 4 is still available on spectrometers with TopSpin versions 2 or newer, if you care to use it.

4. Type gradshim to enter the gradient shimming routine. In the Gradient Shimming window that pops up check that the probe type listed under the "Current Probe" heading is correct, check with your spectrometer administrator if you are in doubt. Next, enter, check, and/or select the following (as in Figure 6.1):

Shimming Method:	1D (1D gradient shimming using ¹H observe)
DISK:	/opt/topspin (or whatever directory path is used with your spectrometer)
USER:	your user login id
FILENAME:	defltld1h (under heading "Iteration Control File")

5. Click "Start Gradient Shimming" to start the process. After the Z-axis gradient shimming completes it will display the fieldmap it obtained for the sample.
   Note: The gradient shimming routine requires that the shim gradient profiles have been determined correctly for your spectrometer/probe/nucleus combination. This is referred to as "shim mapping." If this has not been done, then gradient shimming will not work. If you want to carry out this process (with the permission of your spectrometer administrator) it is presented in Appendix 6.1. Otherwise, just click the "Exit" button in the Gradient Shimming window to terminate the routine. In this case you will need to shim the samples manually.

6. Make sure that your ¹H dataset from step 2 is loaded, you may have to load it manually after gradient shimming completes. Now reacquire the ¹H spectrum. Using the multiple display mode (.md), compare this spectrum with the first spectrum you obtained which was copied to procno 2; use the command rep 2 (read procno 2) in the

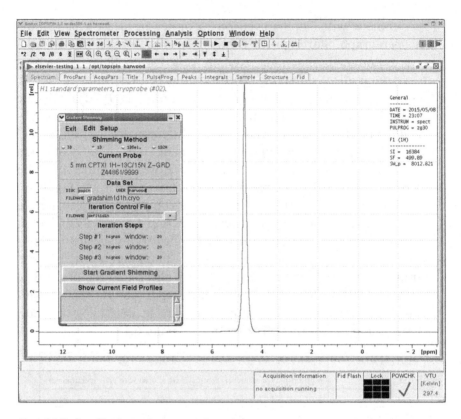

Figure 6.1 Gradient shimming routine, parameter input window.

multiple display mode to do this. Print out a region of the spectrum for comparison.

7. Duplicate to a new expno. Set the chemical shift of the H_2O peak to 4.7 ppm and then put the transmitter frequency (*O1*) directly in the center of the H_2O peak. Rerun the spectrum. Make a mental note of the intensity of the water line compared to the protein signals.

8. Duplicate to a new expno. Set or confirm that *PULPROG* = zg, then find the ^1H 90° pulse calibration as in Chapter 1, Part 3, step 3. Use this calibrated pulse for all the lysozyme experiments run in this laboratory. You may use the `paropt` au program to automate the pulse calibration process. Due to the presence of salt in the sample, the p90 (hence the p360) is expected to be longer than the default calibration value. Note that for *P1* durations approaching the p90 or p270 value, you may see some distortion in the spectrum from the intense H_2O signal, specifically, the H_2O signal may show some dispersive components on its edges.

Part 2—Lysozyme Spectra—Water Suppression Using Presaturation-Based Methods

1. Copy step 8 (above) to a new expno. We will now run the first water-suppression experiment using the simplest method, presaturation. Set the following in ased:

PULPROG	zgpr
TD	8k
NS	16 (or larger multiple of 8, if needed; this applies to all experiments in this Part)
DS	8
P1/PL1	^{1}H 90° pulse calibration from step 7, above
PL9	66 (dB) presaturation power

2. We will need to optimize the value of *O1* to ensure that the transmitter is as accurately centered on-resonance as possible. When this is the case the FID should show no or very little oscillation in its residual water signal. Variations in *O1* of a few Hz or even tenths of Hz can be critical to the performance of presaturation. Type gs to enter the go-setup mode—in this mode the spectrometer acquires one scan of the pulse sequence and repeats it indefinitely, without saving any data. You will see a new FID displayed with every repetition. The gs-Mode window will show a box with various tabs to allow you to select a parameter to optimize (Figure 6.2). Click the "Offset" tab and a slider will show up to allow you to adjust the *O1* parameter while you observe the FID (there is only one offset, *O1*, in the presaturation experiment). Change the "sensitivity" setting to 1. For this experiment we need to adjust the value of *O1* to get the FID to show no oscillations, just a smooth, or nearly smooth, decay. This is easiest to see when using the FID display which shows the two channels separately (the "unshuffled" display—green box in figure). The integrated FID area is displayed in the acquisition status bar at the bottom of the window, and the value is updated with every scan. The integrated FID area will be at a minimum when the water suppression is optimized. When you have obtained the optimum *O1*, click "Save" and "Stop" to save the optimized *O1* value and to halt the gs-operation.

3. You may also adjust the *PL9* value to reduce the residual water signal (up to a minimum attenuation of 51 dB—check with your spectrometer administrator for the specific limit on the

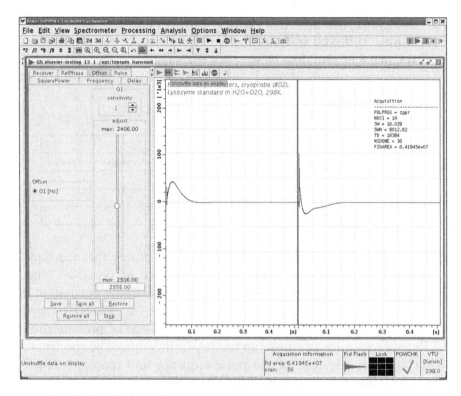

Figure 6.2 gs-Mode window showing O1 *selection and (near) on-resonance FID. (For interpretation of the references to color in the text, the reader is referred to the web version of this book.)*

spectrometer and probe you are using) as much as possible. This is best assessed by running the experiment and looking at the resulting spectrum, then repeating the process with a new *PL9* value 3 dB lower than the previous value (e.g., 66→63→60, etc.). Reductions in the residual water signal will be apparent. Do not decrease the attenuation so much so that distortions arise in the spectrum, or that the sample temperature or the lock signal become unstable.

4. Finally, increase the value of *RG* as much as possible without triggering error messages from the spectrometer or causing distortions in the spectrum. Try to obtain an *RG* value of ca. 32 or greater. This will provide better dynamic range for the acquisition. When you are satisfied that you have obtained the best spectrum possible, obtain your final, fully optimized spectrum.

5. Duplicate to a new expno. We will now use the 1D-NOESY pulse sequence to try to improve the water suppression further. Set the following:

PULPROG	noesypr1d
D8	1 ms (millisecond)
D11	30 ms

All other parameters are the same. Run the new experiment.

6. Copy to a new expno. We will now use the presat180 scheme (reference 73 in Chapter 10 of the Claridge book) and compare the resulting spectrum quality with the previous two spectra. If your spectrometer does not have this pulse sequence installed, you will have to set it up yourself (with your spectrometer administrator's permission). If this is the case please see Appendix 6.2 for the pulse sequence and related instructions. Then, set the following:

PULPROG	presat180
D16	200 µs (microseconds)
P11	500 µs (180° shaped pulse)
PL0	120 (dB)
SP1	(PL1) + (10) (dB, shaped pulse power—must be optimized manually)
SPNAM1	Crp60,0.5,20.1 (shapefile for pulse)
GPNAM1	SINE.50 (gradient shaped pulse file)
GPZ1	5 (%)
P16	500 µs (gradient pulse time)

7. The value of *PL1* + 10 dB given above for *SP1* is just a starting point; for example, if your default *PL1* is 0 dB, start with +10 dB for *SP1*. Now we will optimize the value of *SP1*, but this time we have to actually run the experiment with at least 4 scans (or a multiple thereof) to assess the correct value for *SP1*. To carry out the optimization, set a starting value of *SP1* (e.g., *PL1* + 10 dB) and acquire and process the data. Then, change the *SP1* value by 1 dB and repeat the process. Compare the results to the prior results. You are aiming to improve the water-peak suppression while

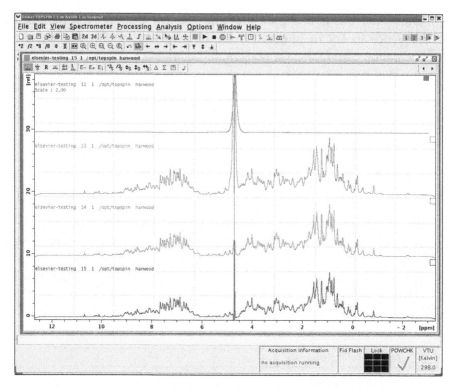

Figure 6.3 Multiple-spectra display of lysozyme / H_2O spectra obtained using presaturation-based water-suppression methods at 500 MHz. Uppermost spectrum is with no water suppression, remaining spectra upper to lower are in the order of the text.

maintaining the intensities of the non-H_2O peaks in the spectrum. It will take some experimentation to find the optimum *SP1* value; it may be larger or smaller (in the absolute sense, i.e., + or −) than the *PL1* + 10 dB starting point we suggest. In the authors' experience, an *SP1* value corresponding to a p90 of about 36 μs (approximately equal to the TOCSY 90° spinlock pulse power (*PL10*) value used in Chapter 4, Part 3, step 2, see also Note 2 for that step) will give acceptable results.

8. After optimizing *SP1*, run the optimized experiment.
9. After obtaining all the data, use the multiple display mode (.md) to compare the three presaturation experiments (steps 2, 3 and 5). Note any differences in the residual water signal intensity and width (Figure 6.3).

Part 3—Lysozyme Spectra—Water Suppression Using "WATERGATE"-Based Methods

1. Copy the last experiment to a new expno. We will now run the first experiment using the simplest WATERGATE method. Set the following:

PULPROG	zggpwg
P11	1200 μs (90° selective pulse)
SP1	power level for P11, ca. 40 (dB, check using edprosol)
SPNAM1	sinc1.1000
GPNAM1	SINE.100
GPZ1	20
P16	1 ms

2. Now optimize the value of *SP1*, using gs as before, looking for the value which gives the minimum FID intensity. *RG* may need to be reduced if the estimate for *SP1* was off.

3. After optimizing *SP1*, run the experiment.

4. Copy to a new expno. We will now use another variant of WATERGATE which uses the 3-9-19 sequence. Set the following:

PULPROG	p3919gp
D19	62, 80, or 100 μs (delay for binomial excitation, these values for 800, 600 or 500 MHz spectrometers, respectively)
P0	Same as *P1*
P27/PL18	Use same values as for *P1/PL1*

All other parameters are the same. Rerun.

5. Copy to a new expno. We will now use a variant of the WATERGATE sequence which uses the w5 excitation/suppression sequence. Set the following:

PULPROG	zggpw5
GPZ1	34
GPZ2	22

All other parameters are the same. Rerun.

6. After obtaining the data, enter the multiple display mode (.md) to compare the three WATERGATE experiments (steps 3, 4, and 5). Note any differences in the residual water signal intensity and width.

Part 4—Lysozyme Spectra—2D NOESY with WATERGATE Water Suppression

1. Copy the ^{1}H WATERGATE w5 spectrum (Part 3, step 5) to a new expno. Set the following in eda:

PULPROG	noesygpphw5
1-2-3	$1 \to 2$ (change 1D–2D)
FnMODE	States-TPPI
TD	2K (F2), 256 (F1)
ND_010	1
SW (Hz)	F2 and F1 both the same as the ^{1}H presaturation spectrum

Note: **ND_010 must be set before the SW values are entered!**

2. Now use ased and enter the following:

NS	8 (depending on your spectrometer's ^{1}H sensitivity, multiples of 8 may be needed)
RG	2 times current value
D8	70 ms (mixing time)
D16	200 µs
D19	62, 80, or 100 µs (see Part 3, step 4)
P27/PL18	Use same values as for *P1/PL1*
PL9	78 (dB, only a low power level is needed for presaturation with WATERGATE in use)
GPNAM1,2	SINE.100
GPZ1	34
GPZ2	22
P16	500 µs (gradient pulse)

3. Now go into edp or the ProcPars tab and set the following processing parameters:

SI	2K (F2), 1K (F1)
SR	F1 value same as F2
WDW	QSINE (both dimensions)
SSB	2 (both dimensions)
PHC0 (F1)	90 (phase corrections in F1)
PHC1 (F1)	−180
PH_mod	pk (both dimensions)
BC_mod	qpol (F2)
FCOR	1 (both dimensions—weighting factor for first data points)
MC2	States-TPPI

4. Run this experiment with zg. This will take about 1.25 h with 8 scans per increment.

5. To set the phase corrections in F2, do rser 1, process and phase the 1D spectrum as was done in Chapter 4, Part 9, step 2, except here we will phase the spectrum to show positive peaks (in Chapter 4 we phased it to be inverted—strictly speaking the NOESY here should be inverted as well but it is conventional in protein NMR to phase the NOESY positive). Save the phase corrections to the 2D dataset.

6. Process with xfb. The spectrum should appear to be phased nearly correctly with all peaks positive. This large protein molecule will have nOe peaks with the same phase as the diagonal. Use the 2D phasing routine to touch up the phasing if necessary.

7. The bas command may be used to flatten the 2D baseplane, if needed (Figure 6.4).

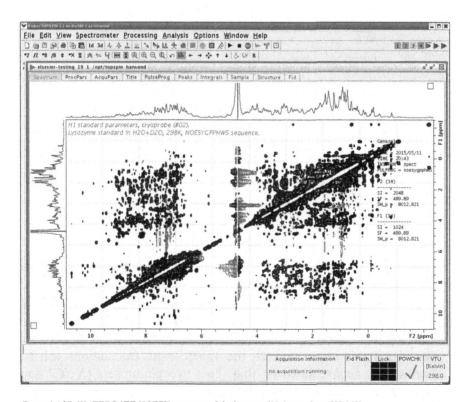

Figure 6.4 2D WATERGATE-NOESY spectrum of the lysozyme/H₂O sample at 500 MHz.

Part 5—Ubiquitin Spectra—Shimming, Pulse Calibration and 1D ^{1}H Spectrum

1. Put in the ubiquitin sample, lock it (use the solvent "$H_2O + D_2O$"), and tune the probe for ^{1}H. Follow the steps from Part 1 of this laboratory to get the initial spectrum (including gradient or manual shimming) and do the ^{1}H pulse calibration. Then use the noesypr1d or the zgpww5 (your choice, make sure to set up everything correctly) pulse sequence to obtain the 1D ^{1}H spectrum with water suppression. The easiest way to do this is to duplicate the experiment run earlier in this Chapter to a new experiment, and update the ^{1}H p90 pulse calibrations before running it on the ubiquitin sample. This serves as a quick evaluation of the sample quality.

Part 6—Ubiquitin 2D ^{13}C HSQC (Heteronuclear Single-Quantum) Experiment Setup and Acquisition

1. We will first set up the ^{1}H–^{13}C HSQC experiment. Copy the 1D spectrum you obtained in Part 5 to a new experiment. In the new experiment set the following in eda:

PULPROG	hsqcetgpsi
1-2-3	$1 \rightarrow 2$
FnMODE	Echo-Antiecho
TD	2K (F2), 256 (F1)
NUC2	13C, then "save," cf. Chapter 2, Part 2, step 3 for the specific steps for your software version; this step sets up the decoupling (F2) channel properly
NUC1 (F1)	13C—this step is important!—sets up the frequency reference in F1
O2P	70 (standard value for the center of the ^{13}C spectrum window)
ND_010	2
SW (F1)	ca. 150 ppm (standard value for the ^{13}C spectrum window)

Note: **ND_010 and NUC1 (F1) must be set before the SW (F1) value is entered!** The *NUC2* entry sets up the decoupling (i.e., the second RF) channel for the correct nucleus (in this case ^{13}C). The *NUC1 (F1)* entry ensures that the software knows that the nucleus/frequency to be used for referencing the frequency calculations in F1 is ^{13}C. If this is not correct, the sweep width calculations will be wrong. Be aware, however, that these two parameters (*NUC2* and *NUC1 (F1)*) are not always the same.

2. In `ased` set the following:

NS	2 or 4 (depending on your instrument's ^{1}H sensitivity)
DS	16
RG	64–1k, depending on the spectrometer, or set using `rga`
CNST2	135 (Hz, the one-bond J_{C-H}—optimized for aliphatic CH)
D1	2
D16	200 u
D24	926 u (1/8 J)
ZGOPTNS	Leave this field blank (flag for enabling ^{15}N decoupling which we will not use)
P1/PL1	^{1}H p90 calibration done in Part 5, above
P28	1 m
P3/PL2	^{13}C p90 calibration (from Chapter 1 or from `edprosol`)
CPDPRG2	garp (CPD sequence for broadband decoupling of ^{13}C)
PCPD2	50–70 (microseconds) (for 800–500 MHz spectrometers)
PL12	ca. 12 dB (power level for 50–70 μs ^{13}C p90—check using `edprosol`)
GPNAM1,2	SINE.100
GPZ1	80 (%)
GPZ2	20.1
P16	1 m

3. In `edp` set the following:

SI	2K in both dimensions
SF (F1)	Set to the BF2 for ^{13}C
WDW	QSINE (both dimensions)
SSB	2 (F2), 3 (F1)
PHC0	0 (both)
PHC1	0 (both)
PH_mod	pk (both)
BC_mod	qpol (F2), no (F1)
FCOR	0.5 (both)
MC2	echo-antiecho

4. Check the probe tuning on ^{13}C.

5. *Optional*: For completeness, the ^{13}C p90 calibration should be checked using this sample prior to acquiring the HSQC spectrum. However, even though the sample is ^{13}C-labeled the low concentration usually precludes direct observation of the ^{13}C signal on most spectrometers. Appendix 6.3 gives the procedure for carrying out this operation.

However, if your spectrometer can directly observe the ^{13}C spectrum of this sample, you may use the direct-observe method to check this calibration (cf. Chapter 1, Part 4).

6. Run this experiment with zg. This will take about 20 min with 2 scans per increment.

Part 7—Ubiquitin 2D ^{15}N HSQC Experiment Setup and Acquisition

1. Copy the ^{1}H–^{13}C HSQC dataset to the next expno. We will now set up the ^{15}N HSQC using the ^{13}C HSQC as a template. In the new experiment set the following in eda:

TD	64 (F1)
NUC2	15N, then "save," see notes in Part 6, above
NUC1 (F1)	15N (see notes in Part 6)
O2P	117 (typical value for the center of the ^{15}N spectrum window)
ND_010	2
SW (F1)	ca. 40 ppm (normal value for the ^{15}N spectrum window)

Note: **ND_010 and NUC1 (F1) must be set before the SW (F1) value is entered!**

2. In ased set the following:

NS	4 (or a multiple of 4, depending on your spectrometer's ^{1}H sensitivity)
DS	16
RG	64–1k, depending on the spectrometer, or set using rga
CNST2	92 (Hz, the one-bond J(NH))
D24	2.7 ms (1/4J—optimized for NH)
ZGOPTNS	Leave this field blank (flag for enabling ^{13}C decoupling)
P3/PL2	^{15}N p90 calibration (from your spectrometer administrator or edprosol)
PCPD2	180–220 µs (for 800–500 MHz spectrometers)
PL12	ca. 10 dB (power level for 180–220 µs ^{15}N p90—check as for *P3/PL2* above)
GPZ2	8.1

3. Type edasp and check the routing of the ^{15}N FCU to its amplifier. This will need to be checked with your spectrometer administrator, since this can be different for different hardware configurations. We show examples of two different common routing possibilities in Appendix 6.4.

4. In `edp` set the following:

SI	512 (F1)
SF (F1)	Set to the BF1 for ¹⁵N

5. Check or carry out the probe tuning for ¹⁵N.
6. *Optional*: For completeness, the ¹⁵N p90 calibration should be checked prior to acquiring the HSQC spectrum. The procedure in Appendix 6.3 can also be used for this operation.
7. Run this experiment with `zg`. This will take about 20 min with 2 scans per increment.

Part 8—Ubiquitin 2D HSQC Experiments Data Processing

1. Go to the first ubiquitin HSQC dataset.
2. Use `xfb` to process the HSQC datasets in turn.
3. After transformation, all experiments may need to be phased manually in both dimensions. When carrying out the 2D phasing, phase

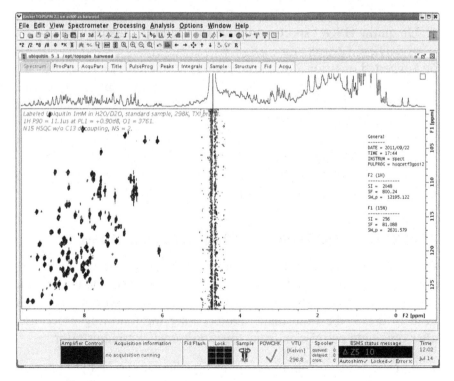

Figure 6.5 2D ¹H–¹⁵N HSQC spectrum of the ubiquitin/H₂O sample at 500 MHz.

the F2 dimension (rows, R) first, before attempting the F1 (columns, C) dimension. Iterate through the F2, F1 phasing sequence if necessary.
4. Set the contour levels in each experiment. Normally these spectra are not displayed or plotted with projections, but you may include an F2 (^{1}H) projection if you wish. Note that since the 2D spectra are not decoupled the cross-peaks may not align with those of the 1D ^{1}H spectrum precisely (Figure 6.5).

SUMMARY

Your lab report should include the following:

1. Printout of the lysozyme sample spectrum with the H_2O line at full intensity (Part 1, step 2).
2. Lysozyme water-suppression spectra from Parts 2 and 3, six spectra in all. Plot each with the same region and vertical scaling to try to compare the signal intensities to residual H_2O.
3. Printout of the lysozyme NOESY spectrum with correct projections.
4. Ubiquitin noesypr1d or zgpww5 spectrum from Part 5.
5. Ubiquitin HSQC spectra from Parts 6 and 7. Ensure that the F2 projection is correct on each plot.
6. Suggested questions for consideration:
   a. What did you get for the ^{1}H p90 calibration for the lysozyme sample? Is this different from the calibration you obtained for the quinine sample? If so, why?
   b. Which of the water-suppression techniques gave the best result in your experience?
   c. For the HSQC experiments, can you suggest anything that could be done to improve the sensitivity?

APPENDIX 6.1 MAPPING SHIM PROFILES IN GRADSHIM

If the shim gradient profiles have not been set up for the current probe, you will receive a message "No 1D shim map exists for the current probe: probe type" when starting `gradshim`. If this occurs you will need to create the shim maps. Do so using the following procedure, **after** obtaining permission from your spectrometer administrator.

1. Exit the gradient shimming routine by clicking the "Exit" button in the Gradient Shimming window.

2. Save the current shims (for later access if needed) by typing `wsh GSHIM`.

3. Reenter the gradient shimming routine by typing `gradshim`. Check that the probe type listed under the "Current Probe" heading is correct, check with your spectrometer administrator if you are in doubt. Check that "1D" is selected for the "Shimming Method." *Note*: This assumes you are using an H₂O-containing sample for this procedure.
Click on the "Setup" button followed by the "Shim Mapping" selection in the pop-down menu which appears. This will open a new window titled "Shim Mapping."

4. In the Shim Mapping window enter, check, and/or select the following (as in Figure A6.1.1):

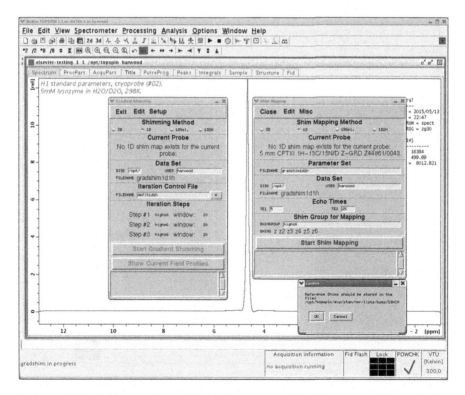

Figure A6.1.1 Input windows of `gradshim` *displayed prior to start of the shim mapping procedure. Note that "highz6" is shown for the SHIMGROUP input.*

Shim Mapping Method:	1D (1D gradient shimming using ^{1}H observe)
FILENAME:	gradshim1d1h (under heading "Parameter Set")
DISK:	/opt/topspin (or whatever data directory path is used with your spectrometer)
USER:	Your user login id
TE1, TE2	5, 25 (under heading "Echo Times")
SHIMGROUP	highz (or highz6, check with your administrator)

5. Click "Start Shim Mapping" to start the process. At this point a small window will pop up telling you that the reference shims should be stored as "GSHIM." Since we already did this in step 2, simply click the "OK" button and the mapping process will start. After the mapping completes it will display the shim profiles it obtains, as shown in Figure A6.1.2.

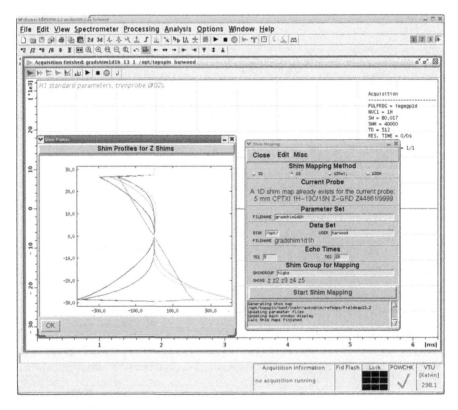

Figure A6.1.2 Shim profiles for Z-shims displayed after completion of shim mapping procedure. Shims are as follows: Z1, red; Z2, blue; Z3, light green; Z4, dark green-gray; Z5, light gray. (For interpretation of the references to color in this figure legend, the reader is referred to the web version of this book.)

APPENDIX 6.2 PRESAT180 PULSE SEQUENCE

If your spectrometer does not have this pulse sequence installed you will need to add it using the `edpul` command as follows.

1. Type `edpul new` to open a new editor window.
2. In the edpul window either paste in (Ctrl-V) or type in the text below starting with the "presat180" line:

```
;presat180
;1D sequence with f1 presaturation
;and adiabatic pulse inversion
;Mo and Raftery JMR (2008) 190:1-6
#include <Avance.incl>
"d12 = 10 u"
"d30 = 0.5"
1 ze
2 30 m
  d12 p19:f1
  d1 cw:f1 ph9
  4 u do:f1
  p16:gp1
  d16 p10:f1
  if "d30 < 1" goto 3
  if "d30 > 1" goto 4
3 p11:sp1 ph29
; d12
  "d30 = 1.5"
  goto 5
4 ;p11:sp2 ph29
  ;d12
  "d30 = 0.5"
5 p16:gp1*-1
  d16 p11:f1
  p1 ph1
  go = 2 ph31
  30 m mc #0 to 2 F0 (zd)
exit
ph1 = 0 0 1 1 2 2 3 3
ph9 = 0
ph29 = {0} *8 {1} *8
ph31 = 0 2 1 3 2 0 3 1
```

```
;pl1: f1 channel - power level for pulse (default)
;pl9: f1 channel - power level for presaturation
;sp1: adiabatic inversion pulse peak power (~8 kHz)
;spnam1: Crp60,0.5,20.1
;p1: f1 channel - 90 degree high power pulse
;p11: adiabatic inversion pulse length [ 500 us]
;p16: gradient pulse length [ 250 us to 500 us]
;d1: relaxation delay > 2 * T1
;d12: delay for power switching      [ 10 usec]
;d16: gradient recovery delay [ 50 us]
;d30: on/off switch for adiabatic pulse
;NS: 8 * n, total number of scans: NS * TD0
;DS: 4 * n
;for z-only gradients:
;gpz1: 5%
;use gradient files:
;gpnam1: SINE.50
```

Make sure that if you type this in that you do not make any mistakes. You can save your input at any time (you will have to use "Save as" as described below the first time and "Save" any subsequent times).

3. Click the "Save as" button and enter presat180.hm in the "New name" text box.
4. Go to the "File" pulldown and select "Exit" to close the editor.

APPENDIX 6.3 X-NUCLEUS PULSE CALIBRATION USING THE gs-Mode

1. Go to the ubiquitin HSQC dataset (Part 6).
2. Start the 2D acquisition. As soon as the first fid is saved, read it using rser 1. Carry out the following:
   a. Type lb and enter 10, do ef, then phase the resulting spectrum with all peaks (excepting the residual H_2O) positive.
   b. When finished phasing, use .s2d or click the save-2D icon to save the phase corrections to the HSQC dataset, followed by .ret or clicking the return icon.
   c. Type to2d to return to the 2D dataset.
3. Stop the current acquisition using the halt command.
4. We will now optimize the pulse calibration for the ^{13}C (or ^{15}N) p90 using the gs-Mode. Type gs to start, and in the gs-Mode window

select "SquarePower" followed by *PL2*. Select the spectrum display mode using the "Execute realtime ft and show spectrum" icon (icon with an FID over a spectrum) above the FID display. Then click the "Realtime ft settings" icon (next to the previous icon) and in the popup window select "PK" for the "Phase correction mode:" option. If necessary for the spectrum quality you may also select an appropriate "Window function:" from the same menu. Click "OK" when finished. You should now be seeing a correctly-phased spectrum updated with each scan.

5. Adjust the variable parameter sensitivity down by a factor of 10. As the gs-Mode is running, adjust the *PL2* value in steps of 0.5 dB to obtain a maximal signal (on average) for the protein (i.e., non-H_2O) peaks of interest in your spectrum when compared with the H_2O peak. Keep in mind that for the ^{13}C HSQC this refers (generally) to the (upfield) aliphatic ^1H peaks, and for the ^{15}N HSQC this refers to the HN (downfield) ^1H peaks. After the optimal value is obtained, stop the gs-operation (cf., Part 2, step 2) and save the new *PL2* value using the popup window which appears upon stopping.

6. If the *PL2* value changes significantly, use edprosol to recalculate the *PL12* value for ^{13}C or ^{15}N garp decoupling.

7. The same procedure with gs can be used to check the ^{15}N p90 prior to running the ^{15}N HSQC experiment.

APPENDIX 6.4 ROUTING FOR THE ^{15}N RF CHANNEL IN THE ^{15}N HSQC EXPERIMENT

The default routing of the ^{15}N channel will depend upon the configuration of your spectrometer.

If your spectrometer has two RF channels, then ^{15}N by default will be on F2. In this case there is no choice to be made for the routing, you will simply have to tune the appropriate coil of the probe to ^{15}N before you proceed with the experiment. This routing arrangement is shown in Figure A6.4.1.

If your spectrometer has three or more RF channels, the experiment can be performed with ^{15}N set up on F2 or F3. The more common default situation for a three- or four-channel spectrometer equipped with a triple- or quadruple-resonance inverse probe (TXI, TBI, or QXI) would be that F2 would be ^{13}C and would be routed to the ^{13}C

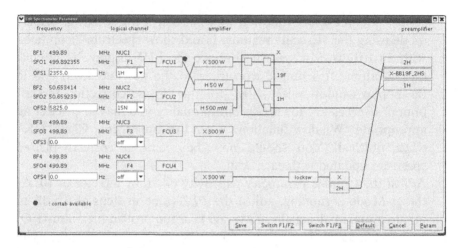

Figure A6.4.1 edasp *display for ^{15}N HSQC experiment representing a two-channel spectrometer with F1 as ^{1}H and F2 as ^{15}N. This picture is from a four-channel spectrometer; a two-channel spectrometer would show only the upper half of the display encompassing F1 and F2.*

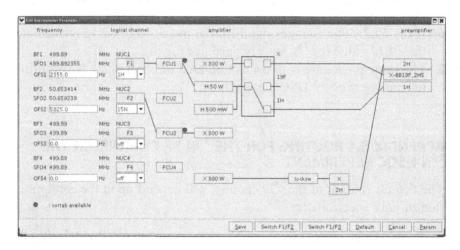

Figure A6.4.2 edasp *display for ^{15}N HSQC experiment for a four-channel spectrometer with the F2 (^{15}N) chan-nel routed via FCU3 to the second X amplifier. The second X amplifier would be connected to the ^{15}N port of an inverse probe.*

port on the probe, and ^{15}N would be F3 and would be routed to the ^{15}N port on the probe. To perform a ^{15}N HSQC experiment with this configuration would require the use of a pulse program which employs F3 for the ^{15}N pulses. The F3-using pulse program analogous to hsqcetgpsi we used in Part 4, above, would be hsqcetf3gpsi.

However, we are using the F2-based pulse program in order to allow the same write-up to be used with (almost) any spectrometer configuration. In this case we will need to route the ^{15}N channel in an appropriate fashion to allow the F2-based pulse program to work with a three- or more channel spectrometer. One way to do this is shown in Figure A6.4.2. Route the output from the "F2" box to the input side of the "FCU3" box, and connect the output of the "FCU3" box to the second X amplifier (by this we mean the second in the listing from the top of the display). This amplifier will typically be routed to the ^{15}N port on an inverse probe.

CHAPTER 7

Selective Experiments Using Shaped Pulses

OVERVIEW

In this chapter's work we will do some computer simulations to investigate the excitation profiles of some shaped RF pulses. We will then apply this to the execution of the selective TOCSY and NOESY experiments. Each of these we will run using two different variants of the pulse sequence—one using a selective 270° pulse and one using a selective 180° pulse combined with field gradients for peak selection. We will use the quinine sample for these experiments. After this we will investigate the use of NMR for quantitation in unusual situations, our examples for this being the determination of the water content in an octanol sample and octanol content in a water sample using the solvent's protons as the concentration reference. The material covered in this chapter is in Chapters 4, 5, 8, and 10 of the Claridge book.

SAMPLE AND SPECTROMETER REQUIREMENTS

This chapter's exercises will use samples A, I, and J. In order to perform the field-gradient enhanced experiments the spectrometer/probe combination in use must have field gradient hardware. The preparation of samples I and J is discussed in Appendix 7.1.

ACTIVITIES

Part 1—^{1}H Setup

1. Put in the quinine sample. Shim the sample and tune the probe for ^{1}H. Leave sample spinning off.
2. Run the ^{1}H spectrum using your previous parameters (include the broad downfield peak).
3. In a new expno, check the ^{1}H 90° pulse calibration, either manually or using `paropt`.

Practical NMR Spectroscopy Laboratory Guide: Using Bruker Spectrometers.
DOI: http://dx.doi.org/10.1016/B978-0-12-800689-4.00007-0

Part 2—Selective Pulse Calculation and Simulation

1. Type `stdisp` within TopSpin to launch the pulse shape analysis tool. Within the new window, you can explore a little bit by selecting some of the sub-menus under "Shapes" and "Analysis."

2. Click the folder icon in the upper left of the TopSpin window. Then select "Open Shape" → "Gauss1.1000." The default shape size has 1000 points and truncates at 1% level (of the maximum intensity; you can change either of them and see what the impact would be, or adjust with the icons). On the right side of the window you will see the shape of this pulse: the amplitude plot has a Gaussian shape and the phase is constant (at zero).

3. Select "Analysis" → "Integrate Shape." Update "90 deg hard pulse" length from the pulse calibration done previously. You will define the pulse length for this shaped pulse (i.e., 27,000 µs) and the total rotation angle (90° for excitation). Now click the "update parameters" button.

4. The last row of the "Results" section shows you the power level difference between this shaped pulse (Gaussian, 27000 µs for 90° excitation) and the reference attenuation found from the previous hard-pulse calibration (Part 1) as ∼63 dB (this depends on your hard 90° pulse length, see Q1). So, if your default 90° pulse calibration uses a power level of +1.5 dB, the absolute power level of this shaped pulse should be ∼(+1.5 dB) + (+63.0 dB) = +64.5 dB. Record your choice of shaped pulse, pulse length/angle and power level (Gaussian, 27,000 µs/90°, ∼+64.5 dB)

5. Select "Analysis" → "Simulation" and update the pulse length/angle as 27,000 µs/90°. Click NMR-SIM and you should see the excitation profile. Print the x, y, and z profiles for this shaped pulse.

6. Repeat step 4 and 5 by changing the total rotation angle to 270° and the pulse length to 80,000 µs. Print the x, y, and z profiles. We will check this estimate in Part 3.

Part 3—Selective Pulse Calibration

(*Note*: For all selective experiments please note the chemical shift center frequency of the selective pulse in the printout's title.)

1. We will now calibrate the 80 ms Gaussian 270° selective pulse. This pulse will be used for both selective TOCSY and NOESY.

2. Copy your reference spectrum from Part 1 to a new expno.

3. In ased enter the following:

PULPROG	selzg
NS	1
DS	0
PL0	120 dB
P11	80 ms (length of shaped pulse)
SPNAM1	Gaus1.1000 (filename for Gaussian shaped pulse)
SP1	75 dB—starting point for the 270° pulse calibration

4. Make a note of the current *O1* value (in Hz). Find the offset (*O2*) for the peak to be selected as before (.octo123). The peak at ca. 4.5 ppm is a good choice.

5. Calculate the value of the parameter *SPOFFS1* (shaped-pulse offset) by *O2 − O1 = SPOFFS1*. Note that *SPOFFS1* may be positive or negative.

6. Type spoffs1 and enter the calculated value from step 5, above. Alternatively, you can change *O1* to equal the *O2* value determined above. This will center the shaped pulse (and in this case, the entire spectrum) on resonance for the desired peak; if you choose this method ensure that *SPOFFS1* is 0.

7. Acquire a spectrum with the *SP1* value set to 75 dB. This is the approximate calibration for a Gaussian 90° pulse. You will need to increase the receiver gain using either the rga routine or by manually entering new values of *RG* and observing the resulting spectrum. You will see the selected peak only in the spectrum, but it will be distorted and show positive and negative intensities—regardless of the phasing. This is due to chemical shift evolution during the pulse. We can get a cleaner peak by employing a 270° pulse, which refocuses this chemical shift evolution.

8. To calibrate the 270° pulse we will need to use a higher power level. Lower the numerical value of *SP1* in 3 dB steps (i.e., increase the RF power stepwise by 3 dB) and check the spectrum at each step. At some point (usually around *SP1* = 63 dB) you will see a peak which is all in-phase and should be phasable to be fully negative. For the calibration purpose use the lowest power level (the most-positive value of *SP1*) which gives a clean inverted peak. You also may use paropt to automate the calibration process, initially use 75 dB as the starting value for *SP1* and use −3 dB (note the −sign) as the increment, for seven increments.

9. Plot your selective-excitation spectrum along with a reference spectrum using the same plot limits as shown in Figure 7.1.

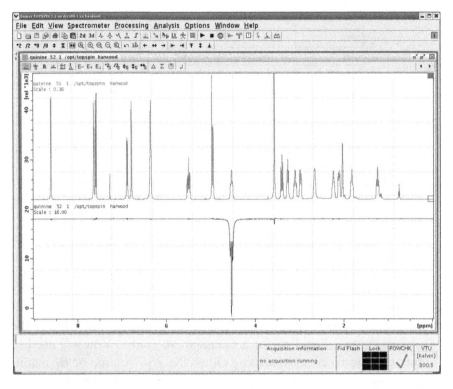

Figure 7.1 Reference quinine spectrum and spectrum with selective excitation using a 270° Gaussian pulse of 80 ms duration centered at ca. 4.56 ppm.

Part 4—Selective TOCSY with 270° Selective Pulse

1. Copy the previous experiment to a new expno.
2. The selective TOCSY experiment will show which peaks are J-coupled to the selectively-irradiated peak, by using a spinlock field applied after the selective pulse for approximately 80 ms. After this a so-called Z-filter is applied to clean up the data, this filter consists of two 90° pulses separated by a random variable delay. This serves to remove any zero-quantum coherence present which would otherwise distort the peaks.

3. In ased enter the following:

PULPROG	selmlzf
NS	32
DS	4
RG	Check for optimal value—keep the same value for all selective experiments
D9	76 ms
D11	30 ms (if requested)
D14	20 µs
TD0	1
P1/PL1	Use the values from Part 1, step 3. This sequence does not employ P1/PL1 but the subsequent ones we run will
P6/PL10	Use the values from the TOCSY experiment in Chapter 4, Part 3
P11/SP1	80 ms and power level determined in Part 3
SPNAM1	Gaus1.1000 (as in Part 3)
P17	2 ms (trim pulse)
PL0	120 dB
VDLIST	Name for VD list file (below)
SPOFFS1	Check the offset is correct for the selected peak (see below, step 5)

4. Now edit a *vd* list for the Z-filter. Type edlist vd to start the editing process and follow the instructions given in Chapter 2, Part 1, step 4. Name for your new file to match what you entered above for *VDLIST*. In the new file enter eight delay values between 0.004 and 0.018 s, in random order, each value on a new line (see the pulse program file for an example of this; **do not** use the delay values given in the example in Chapter 2). Do not enter units, just the numbers alone. Then save the file and exit.

5. Confirm that the *SPOFFS1* corresponds to a peak with apparent J-couplings to other peaks. If so, acquire the data; if not, correct the *SPOFFS1* before acquiring. Alternatively set O1 to put the transmitter on resonance for the selected peak and set *SPOFFS1* to 0 (c.f., Part 3, step 6).

6. Check the receiver gain either manually or with rga before acquiring the data (this step is needed for all the selective experiments performed in this chapter). The spectrum should show the selected peak and then other weaker positive peaks which are J-coupled to the selected one. There will be a bit of subtraction error in the spectrum which will show up as dispersive peaks, but the real peaks should be easy to determine.

7. Duplicate the last dataset to a new one. Pick at least two other peaks for the selective pulse and determine their *SPOFFS1* values,

enter the new *SPOFFS1* values (or change *O1*), and rerun. Use the 2D TOCSY we ran in Chapter 4 as a guide for interesting peaks to select—we especially recommend the peak at ca. 2.0 ppm as a choice. See if your data look reasonable.

8. Print all spectra, use the same plot limits for comparison.

Part 5—Selective TOCSY with 180° Selective Pulse and Field Gradients

1. Copy the previous experiment to a new expno.
2. Go back to Part 2 and use `stdisp` to estimate the power level needed for a 180° Gaussian selective refocusing pulse.
3. In `ased` enter the following:

PULPROG	selmlgp
D16	20 μs
P1/PL1	Use the values from Part 1, step 3 (if not entered already)
P12	80 ms
SP2	Power level for 180° selective pulse—as estimated from `stdisp`
SPNAM2	Gaus1.1000 (same as the previous experiment)
SPOFFS2	Check offset is correct for the selected peak, or change *O1* as discussed previously
GPNAM1	SINE.100
GPZ1	15%
P16	1 ms (gradient pulse)

4. Repeat this selective TOCSY experiment for each of the peaks selected in Part 4. For one of the selected peaks, use the multiple display mode to show both types of selective-TOCSY spectra plotted together. Compare the level of artifacts present in the two spectra.
5. Print all spectra, plus the comparison from step 4, using the same plot limits for comparison.

Part 6—Selective NOESY with 270° Selective Pulse

1. Duplicate one of the selective-TOCSY (270°) spectra from Part 4 to a new expno.
2. In `ased` enter the following:

PULPROG	selnozf
D8	200 ms (nOe mixing time)
P1/PL1	Use the values from Part 1, step 3 (if not entered already)
P11/SP1	80 ms and power level used in Part 4
SPNAM1	Gaus1.1000 (as in Part 4)

3. Decide on a peak to selectively irradiate and determine the correct SPOFFS1 value, or change *O1*.
4. Acquire the data and see if any nOe peaks are present. Duplicate data and then repeat for at least two other selected peaks. Use the 2D NOESY spectrum from Chapter 4 as a guide for picking peaks to irradiate. Other values of *D8* can be tried as well.
5. Print all spectra with descriptive titles.

Part 7—Selective NOESY with 180° Selective Pulse and Field Gradients

1. Duplicate one of the selective-TOCSY (180°) spectra from Part 5 to a new expno.
2. In ased enter the following:

PULPROG	selnogp
D8	200 ms (nOe mixing time)
P12/SP2	From Part 5
SPNAM2	From Part 5
GPZ1	15
GPZ2	40

3. Repeat this selective NOESY experiment for each of the peaks selected in Part 6. For one of the selected peaks, use the multiple display mode to show both types of selective-NOESY spectra plotted together. Compare the level of artifacts present in the two spectra.
4. Print all spectra, plus the comparison from step 6, using the same plot limits for comparison.

Part 8—Quantitation by NMR (qNMR)—Introduction

Modern NMR spectrometers can measure concentrations accurately, especially for proton-containing organic molecules. There are several advantages of qNMR over other quantitation methods. First, a given compound's signature signals (chemical shifts and coupling patterns) can be observed and confirmed as being correct prior to being used for quantitation. Second, the concentration reference can be free of bias: all protons can be detected with same efficiency with correct adjustment of the acquisition parameters. Third, NMR signal generation and detection are linear under proper operating conditions and hence any signal has the potential to be used as a concentration reference, as long as it is reliable and easily measurable. In this section you will use 1D ^{1}H spectra to determine the water content in water-saturated (or nearly-saturated) octanol, and then to determine octanol's solubility in water.

If the experiment's relaxation delay is sufficiently long and/or the excitation angle is very small, then at beginning of each data acquisition, each proton's magnetization is at thermal equilibrium. Under a typical condition for uniform excitation and uniform signal detection, the NMR signal amplitude, if properly integrated, can be expressed as

$$A = A_0 NCI(\theta)\sin\theta$$

where A_0 is a instrument constant if the receiver gain is fixed, C is the concentration of the observed molecule, N is the number of protons within the molecule that contribute to the observed signal, θ is the excitation angle, and $I(\theta)$ is a value close to unity that has a small dependence on θ due to RF inhomogeneity.

For reference and sample protons that are observed within the same experimental acquisition, above equation leads to

$$A_s/A_r = N_s C_s/N_r C_r$$

where the subscript (r) refers to the reference protons and the subscript (s) refers to the sample protons.

Part 9a—Quantitation by NMR (qNMR)—Sample I

1. Put the water-saturated octanol sample I in the magnet. Set the sample temperature to 295 K using edte. Do not attempt to lock the sample. Turn the lock sweep function off.
2. Read the standard ^{1}H parameters, then in ased set the following:

PULPROG	zg
TD	16K
NS	8
DS	4
SW	16 ppm
RG	1
D1	5 s
P1	0.2–2 μs (depending on your spectrometer's ^{1}H sensitivity)
PL1	Normal value for your spectrometer

3. In edp set the following:

SI	16K
LB	0.5

Acquire the spectrum, then set the *O1P* to the approximate center of the spectrum (4 ppm). Check again for the correct receiver gain using rga. If rga gives a value larger than 1, we recommend adjusting *RG* back to ca. one-half of the value obtained from rga. Acquire the spectrum and check to see if the *O1* and/or *SW* parameters need to be optimized. If so, optimize as necessary and repeat the data acquisition.

4. Tune the probe. Then, gradient shim using ^{1}H. If TopShim is available, use the following construct to execute TopShim in the correct way for this particular sample (enter spaces exactly as shown):

topshim 1h lockoff o1p=4 rga

If you are using gradient shimming, be sure to select ^{1}H observation for the shimming, as in Chapter 6, Part 1, step 4, then carry out gradient shimming.

After topshim or gradient shimming completes skip to step 6.

5. If gradient shimming is not available the sample may be shimmed manually using the gs-mode. To do this, first do rga to set the receiver gain. Reduce *D1* to 2 s while running gs then reset it to 5 s after exiting the gs operation. Type gs to start the spectrometer repetitively acquiring one scan. Set up the gs-mode to display the phased spectrum as was done in Appendix 6.3, Step 4. Then, adjust the Z1, Z2, and Z3 shim values while observing the lineshapes and intensities of the peaks in the spectrum. Improving the shimming will result in sharper, more symmetrical, and more intense peaks. To stop the gs operation click the stop-sign icon above the acquisition window or type stop.

Note 1: If the signal intensity is too high with an *RG* of 1, reduce the *P1* duration or the power level *PL1* (i.e., make *PL1* more positive) to obtain a reasonable signal intensity.

Note 2: With a sample as concentrated as this one, radiation damping is likely to cause the ^{1}H signals' line-widths to broaden significantly from normal. It is difficult, and also not necessary, to achieve as good shimming as you usually would for a typical dilute small-molecule sample.

6. Check again for the correct receiver gain using rga. If rga gives a value larger than 1, we recommend adjusting *RG* back to ca. one-half of the value obtained from rga. Acquire the spectrum and check to see if the *O1* and/or *SW* parameters need to be optimized. If so, optimize as necessary and repeat the data acquisition.

7. Process the data and do baseline correction (abs or manually) across the entire spectrum.

8. The most-upfield signal (assigned to the octanol methyl group) chemical shift can be referenced to 0.85 ppm. The two most-downfield signals are assigned to the octanol −OH and water −OH. Integrate all the ^{1}H signals (some of the overlapping methylene groups can integrated together; see Figure 7.2).

9. Reduce the excitation pulse duration by 50%, and repeat steps 4−7. Make sure that neither the lineshape nor the integration ratios change, since radiation damping, depending on the

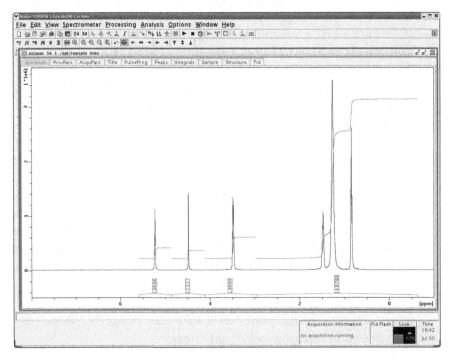

Figure 7.2 1D ^{1}H spectrum at 500 MHz of a fresh water-saturated 1-octanol sample made at room temperature. From downfield to upfield, integrations are from octanol −OH (1), water, 1-CH$_2$ (2) and the remaining octanol aliphatic protons (15). As an internal quality check, the integration ratio of octanol −OH and 1-CH$_2$ is very close to the stoichiometric ratio, while the ratio of the octanol −OH to other combined −CH protons (1:14.88) is about 0.8% lower than predicted. Using the density of water-saturated octanol as 0.8326 g/mL, and the total octanol proton integration vs. that of water (0.738), the water content is calculated as 40.4 mg/mL.

concentration, magnetic field and probe Q factor, can potentially make the integration less reliable for larger excitation angles. Repeat this step if necessary by reducing the pulse length until the integration ratios remain constant. If higher sensitivity is needed, adjust *NS* and *RG* to improve signal intensity.

10. Repeat the data acquisition and processing two more times in different expnos.
11. Calculate the water concentration, based on the mixture's known density of 0.8326 g/mL and the molecular weights of 130.23 for octanol and 18.015 for water.
12. Plot a representative spectrum and include the calculated water content in the title text.

Part 9b—Quantitation by NMR (qNMR)—Sample J

1. Put the octanol-in-water sample, sample J, in the magnet. Set the sample temperature to 295 K using edte. Do not attempt to lock the sample. Turn the lock sweep function off.
2. Repeat the steps 2−7 from part 9a. *NS* can be increased as necessary so that the octanol signal can be clearly observed along with the water.
3. Carefully manually phase the ^{1}H spectrum so that the baseline is as flat as possible. Do baseline correction (abs or manually) across the entire spectrum. The water signal chemical shift can be referenced to ca. 4.8 ppm.
4. Zoom into the octanol methyl region (0.5−2 ppm) and, if necessary, apply further manual first order baseline correction. For the octanol signal, a good baseline is essential for the quality of the weaker signal's integration.
5. Integrate the water and octanol −CH3 group signals. For ease of reporting, the water integration can be normalized to 1,000,000.
6. Reduce the excitation pulse duration by 50%, and repeat steps 2−5. Make sure that neither the lineshape nor integration ratios change. Repeat this step if necessary by reducing the pulse length until the integration ratios remain constant. If higher sensitivity is needed, adjust *NS* and *RG* to improve signal intensity.
7. Repeat the data acquisition and processing two more times in different expnos (Figure 7.3).

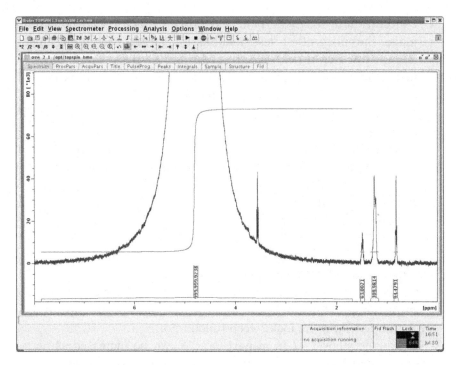

Figure 7.3 1D ^1H spectrum at 500 MHz of a sample of octanol in water made at room temperature. Though with limited sensitivity (constrained by the unsuppressed water signal intensity), the integration ratios of the octanol protons (<2 ppm) are still close to the expected value (2:10:3). If we assume the water ^1H concentration of 111 M, using the water integration (1,000,000) and the octanol methyl integration the octanol concentration is calculated as 3.39 mM or 0.45 g/L (0.46 g/L from CRC Handbook of Chemistry and Physics. 92nd Edition. Editor-in-Chief: W.M. Haynes. 2011–2012. CRC Press, Boca Raton, FL, USA).

8. Calculate the octanol concentration using a density of 0.9982 g/mL, water MW 18.015, octanol MW 130.23, and water ^1H as well as the octanol methyl ^1H integration.
9. Plot a representative spectrum and include the calculated octanol content in the title text.

SUMMARY

Include the following with your lab report:

1. Results and printouts from simulations (Part 2, steps 4, 5, and 6).
2. Results from selective pulse calibration and the multiple display printout (Part 3).
3. All selective TOCSY spectra and one multiple display comparison (Part 5, step 4).

4. All selective NOESY spectra and one multiple display comparison (Part 7, step 3).
5. Plots of representative spectra of the two quantitation samples.
6. Report your average calculated water concentration in octanol from Part 9a, and the standard deviation based on the three separate acquisitions.
7. Report your average calculated octanol concentration in water from Part 9b, and the standard deviation based on the three separate acquisitions.
8. Suggested questions for consideration:
   a. In Part 2, step 4, why does the required power level depend on the 90° pulse length, if the shaped pulse length is fixed?
   b. In Part 2, step 5 and 6, which do you think would be the more-selective excitation pulse (Gauss 10 ms/90° vs. Gauss 30 ms/270°). Why?
   c. Do you find any significant difference in between calculated power level and the experimental value for the 270° Gaussian pulse? Can you give an explanation for the finding?
   d. Did you find any significant differences between results obtained for the selective-excitation experiments compared to their 2D counterparts we obtained in Chapter 4?
   e. The water content in water-saturated octanol has been reported with different values. Based on Karl–Fischer titration, a recent literature value of the water content in water-saturated 1-octanol at room temperature is 40.5 mg/mL (Lang, B. E., *J. Chem. Eng. Data*, **2012**, 57 (8), 2221–2226). How does your value compare with that cited value? Do you think your result, in terms of precision, is better or worse than the literature, and why?
   f. It has been observed by the authors that analysis of a water-saturated octanol sample after prolonged storage in a sealed NMR tube shows a lower water content compared to a freshly-made sample. Can you give a possible explanation of this phenomenon?

APPENDIX 7.1 QUANTITATION SAMPLES I AND J PREPARATION

Sample I can be made up by vigorously mixing 60 mg water in 1 mL of 1-octanol for 10 min at room temperature. After that the water and octanol layers can be separated best by centrifugation, but it will separate on its own if left in a stoppered test tube with minimal head space for a few days. Be sure to use the octanol (upper) layer for the NMR experiment.

Sample J can be made by mixing 2 mg of 1-octanol in 1 mL of water. The rest of the preparation is the same as above, except that the sample will only have one visible layer.

CHAPTER 8

Diffusion Measurements and DOSY (Diffusion Ordered SpectroscopY)

OVERVIEW

Modern NMR spectrometers can readily measure self-diffusion coefficients for molecules in solution with effective MW ranging from 18 Da (water) to more than 10,000 Da. In this chapter you will acquire DOSY (Diffusion Ordered SpectroscopY) spectra for residual water in a D_2O sample and then a test sample (M). The purpose of the water DOSY is to check the gradient calibration we performed with the imaging method in Chapter 5. The test sample M shows that molecules of different sizes in a mixture can be "separated" spectroscopically according to each component's self-diffusion coefficient. This material is covered in Chapter 9 of the Claridge book.

A block pulse sequence diagram for a stimulated echo diffusion experiment with bipolar field gradients is shown in Figure 8.1.

The observed signal intensity (I_0) is given as

$$I = I_0 e^{-D\gamma^2 g^2 \delta^2} \left(\Delta - \frac{\delta}{3} - \frac{\tau}{2} \right)$$

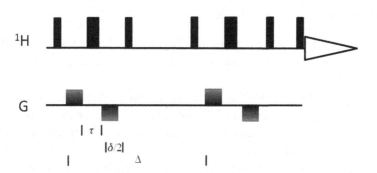

Figure 8.1 Stimulated echo pulse sequence for NMR diffusion measurement employing bipolar field gradients.

Practical NMR Spectroscopy Laboratory Guide: Using Bruker Spectrometers.
DOI: http://dx.doi.org/10.1016/B978-0-12-800689-4.00008-2

where I is the observed signal intensity, D is the diffusion coefficient, g is the gradient, γ is the gyromagnetic ratio, and Δ, δ and τ are diffusion time, length of gradient, and delay between gradients, respectively. A natural logarithm plot of I against $\gamma^2 g^2 \delta^2 (\Delta - \delta/3 - \tau/2)$ gives a negative slope as $-D$.

An important component of diffusion measurements is the temperature of the sample, since the self-diffusion coefficient is sensitive to temperature. Before we start any diffusion measurements we will go through the process of calibrating your spectrometer's variable sample temperature (VT) control system. This is a useful exercise in general since the sample temperature is relevant to many NMR measurements.

SAMPLE AND SPECTROMETER REQUIREMENTS

This chapter's exercises will use samples K, L, and M. Since we are investigating diffusion using pulsed field gradient experiments the spectrometer and probe must be equipped with the appropriate field gradient hardware. The spectrometer also should be equipped with a VT control system in order to get the best diffusion results and to be able to do the VT calibration experiment, but it is not strictly required to perform the diffusion experiments.

ACTIVITIES

Part 1—VT Calibration

Here we will conduct a sample temperature calibration using the neat methanol sample, which has a calibration range from 240 to 310 K. When calibrating sample-temperatures above 310 K, a sample of pure ethylene glycol is used instead.

1. Put sample K into the magnet. Turn down the lock power and turn off the lock and the lock sweep function. Leave sample spinning off.

   Note: You may carry out the temperature calibration with the sample spinning, but for our diffusion measurements in Parts 2 and 3 we will leave the spinning off. Therefore we will do our calibration now with spinning off.

2. In "edte" window, check under the "correction" tab that to make sure that the slope is 1.0 and the offset is 0 (if changes are to be made, record the prior values).

3. Set a desired target temperature (this is the read temperature). A temperature of ca. 298 K (which is what we will use in Parts 2 and 3) can usually be obtained without the use of any cooling equipment for the probe VT gas. If a significantly higher or lower sample temperature is desired, make sure that the probe is specified for the desired temperature, and that the cooling and heating hardware are available, enabled, and have sufficient power. Adjust the maximum heater power (limit) consistent with the set temperature; for most spectrometers, a limit at or below 10% (of maximum power) will suffice for this experiment. If the final read temperature is unstable, the temperature controller may be tuned using the "Tune" function in `edte`. Check with your spectrometer administrator before attempting this.

4. After steady temperature is achieved, record the gas flow rate, chiller condition (on/off), and heater power. If necessary, e.g., if the sample temperature is unstable, the gas flow rate can be adjusted, but you will need to wait for equilibration after the change. On the authors' spectrometers we normally use a gas flow rate of 535 (occasionally 670) L/h.

5. Acquire a ^{1}H spectrum (1 scan only; with a short $P1$ pulse length of 1 μs or shorter). Since we cannot lock this sample, you will need to shim this sample by using the ^{1}H FID in the gs-mode as described in Chapter 7, Part 9a, step 4. Shim enough so that both the hydroxyl and methyl signals are reasonably sharp and their positions can be accurately read (with an uncertainty of ca. 1 Hz or better).

6. Set appropriate MI and $MAXI$ values so that the `pps` command can pick both the −OH and −CH$_3$ peaks. Record their chemical shift difference (Δ) in ppm and calculate the actual sample temperature by using the equation $T = 403 − 29.53\Delta − 23.87\Delta^2$ (Eqn (3.22) in the Claridge book, 2nd ed.).

7. Repeat steps 2−4 as many times as needed, with different target temperatures. For the diffusion experiments in Parts 2 and 3 we would recommend covering a sample temperature range of 288−303 K using 5-degree (K = °C) steps (or smaller, a minimum of four experiments) for the calibration exercise. If your spectrometer does not have any VT gas cooling equipment, a sample temperature of 288 K might not be attainable. In this case just let the sample temperature equilibrate at its lowest point with the heater power limit set to 0%.

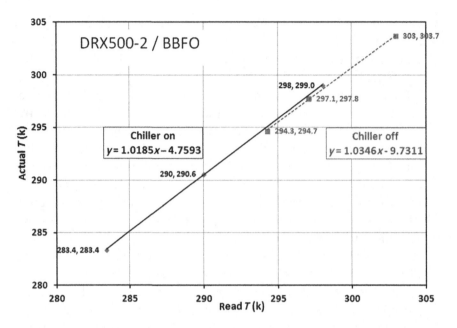

Figure 8.2 VT calibration plots using the neat methanol sample in a 500 MHz spectrometer equipped with a 5-mm BBFO Z-gradient probe for sample-temperatures over the range of ca. 283–304 K. The two different lines represent VT operation with or without the use of a refrigerated chiller to cool the probe VT gas.

8. Plot the actual sample temperature against the read temperature using all the data points you have acquired, as shown in Figure 8.2.

9. Make a note of the relevant conditions for your calibration plot (chiller on/off, spinning on/off, sample and read temperatures, etc.).

Part 2—Diffusion of H$_2$O in D$_2$O

1. Replace sample K in the spectrometer with sample L (if you do not have sample L a sample of D$_2$O may be used, with a small amount of H$_2$O added if desired). After the desired temperature (25°C/298 K recommended) is set and equilibrated, lock, shim, tune the probe for ^{1}H, and acquire the ^{1}H spectrum. Then optimize the observe window and rerun (for samples in D$_2$O/H$_2$O, the transmitter frequency *O1* is frequently placed at the location of the water peak). For the diffusion experiments an acquisition time (*AQ*) of ca. 1–1.5 s is adequate.

2. Copy to a new expno. Set *PULPROG* = zg, then carry out the ^{1}H 90° pulse calibration as in Chapter 1, Part 3, step 3 and make a note of the result. You may use the paropt program to automate the pulse calibration process, if desired.

3. Copy to a new expno (for the DOSY experiment). In the `ased` window set the following:

PULPROG	ledbpgp2s1d
NS	8 (or a multiple of 8 if needed for sensitivity)
DS	4
D1	3 s
D16	200 µs
D20	50 ms
D21	2 ms
P1/PL1	^{1}H 90° pulse calibration from step 2, above
P19	500 µs
P30	2 ms
GPNAME*	SINE.100
GPZ6	+2
GPZ7	−17.13
GPZ8	−13.17

Check that the calculated delays *DELTA1* and *DELTA2* are small positive numbers. If they are not, check for an error or an unrealistic value in one of the input parameters.

4. Set processing parameters $SI = TD/2$ and $LB = 1$. Do `rga`, then acquire and process the spectrum.

5. Copy this to a new experiment, then change *GPZ6* to +95. Acquire and process the spectrum exactly the same way as above but **without** doing `rga`.

6. Compare the spectrum obtained in step 5 with that from step 4. In the step 5 spectrum, the signal(s) of interest should be ca. 10% (or less) of the intensity of the corresponding signal(s) in the step 4 spectrum. If this is not the case, increase *P30* to up to 5 ms and/or increase *D20* (up to the order of T_1). Reduce *D21* if needed to make sure *DELTA1* and *DELTA2* remain small positive values. If the expected diffusion coefficient is small, the gradient pulse shapes (*GPNAME**) may be changed to a smoothed square pulse (GRADREC5m).

7. If you made any parameter changes in step 6, you will need to repeat steps 4−6 again. Check to make sure that the final spectrum has sufficient signal intensity for the peak(s) of interest. Increase the number of scans (in multiples of 8) if needed.

8. Create a new experiment based on your final experiment from step 5. In `eda` set the following:

PULPROG	ledbpgp2s
1–2–3	1→2 (change 1D–2D)
FnMODE	QF (magnitude-mode)
DS	8
TD	32 (F1)
ND_010	1
GPZ6	+100

9. Now use the ProcPars tab or type `edp` and enter the following processing parameters:

SI (F2)	Set equal to TD (F2)/2
SI (F1)	32
WDW (F2)	EM
LB (F2)	1
PH_mod (F2)	pk
PH_mod (F1)	no

10. After the sample has equilibrated at least 10 min at the target temperature, start the data acquisition by `xau dosy 2 95 32 1 y n`. This construct means the following: execute the au program "dosy" with the gradient strength (*GPZ6*) increasing linearly in 32 steps (same as TD(F1) in step 8) from 2% to 95%, to begin immediately and without doing `rga`.

11. Before doing the data processing, check the gradient calibration by using either `gradpar` or `setpre`. Enter the gradient calibration value obtained from the 1D image method used in Chapter 5, Part 1. The gradient calibration has a direct impact on the measured diffusion coefficient. If the gradient calibration is updated, execute `xau dosy restore` and then reprocess the DOSY spectrum.

12. Do `xf2`, followed by `abs2`, `setdiffparm`, and then `eddosy`. Normally the default parameters are sufficient for the DOSY experiment on HDO in D_2O. Click on the 2D DOSY spectrum display within the `eddosy` sub-window. To reprocess the data for any reason, start over from the `xf2` step. NOTES: While the phase corrections are inherited from the previous 1Ds, you may still need to confirm that all FIDs are acquired and phased properly.

The *ABF1* and *ABF2* parameters may need to be adjusted for the abs2 baseline correction to give the optimum result. The *I1max* and *I1min* parameters may be used to ignore certain large or small signals in eddosy.

13. Click on spectrum window within the eddosy sub-window to start DOSY fitting. The 2D DOSY spectrum should appear (if not, click the Spectrum tab), with the ^1H chemical shift for the horizontal axis and the self-diffusion coefficient for the vertical axis. Proton signals from the same molecule should have the same diffusion coefficient.

14. The self-diffusion coefficient of HDO in D_2O at 25°C is 1.90×10^{-9} m²/s. If your DOSY spectrum shows the peak at a significantly different diffusion coefficient than this, adjust the gradient calibration used in step 11. Then repeat steps 12 and 13 so that the observed HDO peak appears at 1.90×10^{-9} m²/s or at $\log D = -8.72$, as shown in Figure 8.3. If the temperature has been calibrated and the experiments are performed properly, the gradient strength calibration should be close to the one obtained from the 1D image method used in Chapter 5, Part 1.

Part 3—DOSY Spectrum of Sample M

1. Replace sample L in the spectrometer with sample M, the SDS/ ATP sample. This sample is made up using 3.6 mg/mL SDS and 2 mM ATP in 50 mM NaPi pH 6.8 D_2O buffer. However, you can

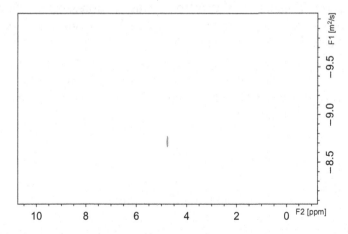

Figure 8.3 DOSY spectrum of HDO in D_2O. With the Z-gradient calibration of 6.0 G/mm, the HDO self-diffusion coefficient is measured in the DOSY spectrum as 1.9×10^{-9} m²/s at 25°C, or at $\log D = -8.72$.

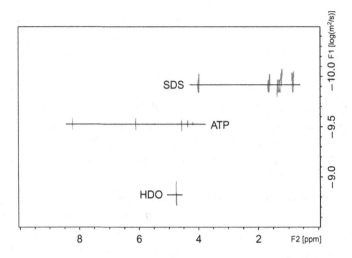

Figure 8.4 DOSY spectrum of 3.6 mg/mL SDS and 2 mg/mL ATP in 50 mM NaPi pH = 6.8 D₂O solution. HOD's self-diffusion coefficient serves as an internal control, with its reading being slightly lower than in pure D₂O solvent, presumably due to presence of SDS, ATP, and buffer.

use your own sample for this step, if desired. After the desired temperature (25°C/298 K recommended) is set and equilibrated, lock, shim, tune the probe for ^{1}H, and acquire the ^{1}H spectrum. Then optimize the observe window and rerun (for samples in D_2O, the transmitter frequency *O1* is frequently placed at water peak). Here again an acquisition time of ca. 1–1.5 s is adequate.

2. Acquire DOSY spectrum as described in Part 2. *Note*: Since both SDS and ATP diffuse slower than water, you may need to increase *P30* (up to 5 ms) and/or *D20* (to 100 ms or more, depending on T_1). In the `eddosy` step, *DISPmax* and *DISPmin* (using the logarithmic scale) can be set to -8.5 and -10.5.

3. You may adjust the various options in the `eddosy` parameter settings. You can toggle the curve fitting mode between peak integrals or intensities, or *I1max*. Figure 8.4 shows the DOSY spectrum of sample M.

SUMMARY

Your lab report should include the following:

1. A plot of a methanol spectrum from Part 1 showing the distance (Δ, in ppm) between the $-OH$ and $-CH_3$ ^{1}H peaks. Include both

the temperature set point and the measured temperature in the title text of the plot.

2. A plot of the temperature calibration data you obtained from part 1.

3. A DOSY plot of HDO in D_2O, including a list of experimental parameters such as field strength, probe, sample details, temperature, pulse sequence, Δ and δ.

4. A DOSY plot of SDS and ATP, with experimental parameters as above.

5. Suggested questions for consideration:

   a. Can you suggest a reason for why methanol's —OH chemical shift is sensitive to temperature?

   b. Explain why the diffusion of HDO in D_2O ($1.9 \times 10^{-9}\,m^2/s$) is slower than H_2O in H_2O (consensus value of $2.3\ 10^{-9}\,m^2/s$) (Table 1 in M. Holz and H. Weingartner, *J. Magn. Reson.* **1991**, *92*, 115–125).

   c. Estimate the average size of SDS micelles using some simple assumptions: reference ATP is monomeric and spheric, volume is proportional to MW, and D is inversely proportional to radius r.

   d. Temperature gradients in the sample can cause convection. Evaluate the impact on the diffusion coefficient measurement due to convection for small and large species (e.g., water and SDS micelles).

   e. Regarding sample M, are you able to observe the diffusion coefficient of HDO in this sample? How does this value compare with your expectations?

Printed in the United States
By Bookmasters